Health and Safety Commission

Work with materials containing asbestos

Control of Asbestos Regulations 2006

Approved Code of Practice and guidance

HSE Books

First published 2006

ISBN 0 7176 6206 3

This Code has been approved by the Health and Safety Commission, with the consent of the Secretary of State. It gives practical advice on how to comply with the law. If you follow the advice you will be doing enough to comply with the law in respect of those specific matters on which the Code gives advice. You may use alternative methods to those set out in the Code in order to comply with the law.

However, the Code has a special legal status. If you are prosecuted for breach of health and safety law, and it is proved that you did not follow the relevant provisions of the Code, you will need to show that you have complied with the law in some other way or a Court will find you at fault.

Contents

Notice of Approval

By virtue of section 16(4) of the Health and Safety at Work etc. Act 1974 ("the 1974 Act"), and with the consent of the Secretary of State for Work and Pensions pursuant to section 16(2) of the 1974 Act, the Health and Safety Commission has on 25 July 2006 approved the Code of Practice entitled *Work with materials containing asbestos* (First edition 2006).

The Code of Practice comes into effect on 13 November 2006.

The Code of Practice gives practical guidance on safely controlling work with asbestos-containing materials. It should be read in conjunction with the Code of Practice entitled *The management of asbestos in non-domestic premises* and other HSE guidance.

By virtue of section 16(5) of the 1974 Act and with the consent of the Secretary of State under that paragraph, the Health and Safety Commission has withdrawn the following Codes of Practice:

> *Work with asbestos insulation, asbestos coating and asbestos board (Fourth edition)* approved by the Commission on 11 November 2002, and
>
> *Work with asbestos which does not normally require a licence (Fourth edition)* approved by the Commission on 11 November 2002,

which both ceased to have effect on 13 November 2006, being the date on which the Control of Asbestos Regulations 2006 came into force.

Signed

SUSAN MAWER
Secretary to the Health and Safety Commission

20 October 2006

Preface

This publication contains the Control of Asbestos Regulations 2006 (the Asbestos Regulations)[1] together with an Approved Code of Practice (ACOP) and additional guidance.

For convenience, the text of the draft Regulations is included in *italic* type, with the accompanying guidance in normal type and the ACOP in **bold** type.

Introduction

1 This ACOP applies to all work with asbestos. It applies in particular to work on, or which disturbs or is liable to disturb, materials containing asbestos, asbestos sampling and laboratory analysis. The clearing of asbestos-contaminated land is not specifically covered by this ACOP. An additional ACOP entitled *The management of asbestos in non-domestic premises*[2] is aimed at those who have repair and maintenance responsibilities for non-domestic premises.

2 Although only a court can give an authoritative interpretation of the law, in considering the application of this ACOP and guidance to people under your control and direction who are self-employed for tax and/or National Insurance purposes, they are likely to be treated as your employees for health and safety purposes. You may therefore need to take appropriate action to protect them. If you are in any doubt about who is responsible for the health and safety of a person working for you, this could be clarified and included in the terms of a contract. However, remember you cannot pass on a legal duty that falls to you under the Health and Safety at Work etc Act 1974 (the HSW Act)[3] by means of a contract and you will still retain duties towards others by virtue of section 3 of the HSW Act. If you intend to employ such workers on the basis that you are not responsible for their health and safety, you should seek legal advice before doing so.

Meaning of 'reasonably practicable'

3 The term 'so far as is reasonably practicable' is used several times in this document, and needs to be clearly understood in the context of these Regulations. It has been interpreted by the courts as allowing economic considerations to be taken into account as one factor with, for example, time or trouble, to be set against the risk. It is reasonably practicable to take measures up to the point where the taking of further measures becomes grossly disproportionate to any residual risk. The greater the risk, the more likely it is that it is reasonable to go to substantial expense, trouble and invention to reduce it. However, if the risk is small, it would not be considered reasonable to go to great expense. Ultimately, the judgement is an objective one based on the health risks and not on the size or financial position of the employer.

Consulting employees and/or safety representatives

4 Proper consultation with those who do the work is crucial in helping to raise awareness of the importance of health and safety. It can make a significant contribution to creating and maintaining a safe and healthy working environment and an effective health and safety culture. In turn, this can benefit the business in making it more efficient by reducing the number of accidents and incidents of work-related ill health.

5 Employers must consult safety representatives appointed by recognised trade unions under the Safety Representatives and Safety Committees Regulations 1977.[4] Employees who are not covered by such representatives must be consulted, either directly or indirectly through elected representatives of employee safety under the Health and Safety (Consultation with Employees) Regulations 1996.[5] More information on an employer's duties under these Regulations is contained in a free leaflet *Consulting employees on health and safety: A guide to the law*.[6] Such consultations allow employees or their representatives to help employers develop training and control measures.

General advice on complying with the Control of Asbestos Regulations 2006

6 Most of the duties in the Asbestos Regulations are placed upon 'an employer', that is, the person who employs the workers who are liable to be exposed to asbestos in the course of their work. Although the Regulations always refer to an employer, regulation 3(1) makes it clear that self-employed people have the same duties towards themselves and others as an employer has towards his or her employees and others.

7 There is an exemption from certain regulatory requirements for particular, specified types of work with asbestos where any worker exposure will only be sporadic and of low intensity and the exposure level is below the control limit (regulation 3(2)). Such work will not require a licence. All other work with asbestos will require a licence (regulation 8); must be notified to the relevant enforcing authority (regulation 9); must have emergency arrangements in place (regulation 15(1)); must have designated asbestos areas (regulation 18); and those working with the asbestos must be subject to medical surveillance and have health records (regulation 22). Some of the guidance in this ACOP is specifically aimed at this more hazardous work and, for convenience, this work will be referred to as licensable work throughout the ACOP.

8 If the control limit for asbestos is exceeded in the working area, this triggers particular requirements including:

(a) immediately informing employees and their representatives (regulation 11(5)(b)(i));

(b) identification of the reasons for the control limit being exceeded and the introduction of appropriate measures to prevent it being exceeded again (regulation 11(5)(b)(ii));

(c) stopping work until adequate measures have been taken to reduce employees' exposure to below the control limit (regulation 11(5)(b)(iii));

(d) a check of the effectiveness of the measures taken to reduce the levels of asbestos in the air by carrying out immediate air monitoring (regulation 11(5)(b)(iv));

(e) the designation of respirator zones; and

(f) the mandatory provision of respiratory protective equipment (regulation 11(3), although such equipment should always be provided if it is reasonably practicable to do so (regulation 11(2)).

9 Where work with asbestos forms part of a larger project there will be a particular need to co-operate with other employers, and there may be other Regulations (such as the Construction (Design and Management) Regulations 1994 (as amended))[7] which must be taken into account. However, the responsibility to ensure compliance with the provisions of the Asbestos Regulations remains with the employer or self-employed person described in paragraph 6.

PART 1 PRELIMINARY

Regulation 1

Regulation 1

Citation and commencement

These Regulations may be cited as the Control of Asbestos Regulations 2006 and shall come into force on 13th November 2006, except regulation 20(4) which shall come into force on 6th April 2007.

Regulation 2

Regulation 2

Interpretation

(1) In these Regulations –

"adequate" means adequate having regard only to the nature and degree of exposure to asbestos, and "adequately" shall be construed accordingly;

"appointed doctor" means a registered medical practitioner appointed for the time being in writing by the Executive for the purpose of these Regulations;

"approved" means approved for the time being in writing by the Health and Safety Commission or the Executive as the case may be;

"asbestos" means the following fibrous silicates –

(a) asbestos actinolite, CAS No 77536-66-4(*);

(b) asbestos grunerite (amosite), CAS No 12172-73-5(*);

(c) asbestos anthophyllite, CAS No 77536-67-5(*);

(d) chrysotile, CAS No 12001-29-5;

(e) crocidolite, CAS No 12001-28-4(*); and

(f) asbestos tremolite, CAS No 77536-68-6(*),

and references to "CAS" followed by a numerical sequence are references to CAS Registry Numbers assigned to chemicals by the Chemical Abstracts Service, a division of the American Chemical Society;

"the control limit" means a concentration of asbestos in the atmosphere when measured in accordance with the 1997 WHO recommended method, or by a method giving equivalent results to that method approved by the Health and Safety Commission, of 0.1 fibres per cubic centimetre of air averaged over a continuous period of 4 hours;

"control measure" means a measure taken to prevent or reduce exposure to asbestos (including the provision of systems of work and supervision, the cleaning of workplaces, premises, plant and equipment, and the provision and use of engineering controls and personal protective equipment);

"emergency services" include –

(a) police, fire, rescue and ambulance services;

(b) Her Majesty's Coastguard;

Regulation

"employment medical adviser" means an employment medical adviser appointed under section 56 of the Health and Safety at Work etc. Act 1974;

"enforcing authority" means the Executive, local authority or Office of Rail Regulation, determined in accordance with the provisions of the Health and Safety (Enforcing Authority) Regulations 1998[a] and the provisions of the Health and Safety (Enforcing Authority for Railways and Other Guided Transport Systems) Regulations 2006;[b]

"the Executive" means the Health and Safety Executive;

"ISO 17020" means European Standard EN ISO/IEC 17020, "General criteria for the operation of various types of bodies performing inspection" as revised or reissued from time to time and accepted by the Comité Européen de Normalisation Electrotechnique (CEN/CENELEC);[c]

"ISO 17025" means European Standard EN ISO/IEC 17025, "General requirements for the competence of testing and calibration laboratories" as revised or reissued from time to time and accepted by the Comité Européen de Normalisation Electrotechnique (CEN/CENELEC);[d]

"medical examination" includes any laboratory tests and X-rays that a relevant doctor may require;

"personal protective equipment" means all equipment (including clothing) which is intended to be worn or held by a person at work and which protects that person against one or more risks to his health, and any addition or accessory designed to meet that objective;

"relevant doctor" means an appointed doctor or an employment medical adviser;

"risk assessment" means the assessment of risk required by regulation 6(1)(a);

"the 1997 WHO recommended method" means the publication "Determination of airborne fibre concentrations. A recommended method, by phase-contrast optical microscopy (membrane filter method), WHO (World Health Organisation), Geneva 1997.[e]

(2) For the purposes of these Regulations, except in accordance with regulation 11(3) and (5), in determining whether an employee is exposed to asbestos or whether the extent of such exposure exceeds the control limit, no account shall be taken of respiratory protective equipment which, for the time being, is being worn by that employee.

(3) A reference to work with asbestos in these Regulations shall include –

(a) work which consists of the removal, repair or disturbance of asbestos or materials containing asbestos;

(b) work which is ancillary to such work; and

(c) supervision of such work and such ancillary work.

(a) S.I. 1998/494, as amended by S.I. 1999/2024, S.I. 1999/3232, S.I. 2002/2675, S.I. 2004/3168 and S.I.2006/557.
(b) S.I. 2006/557.
(c) The most recent version is reference number EN ISO/IEC 17020: 2004, accepted by CEN/CENELEC on 15th July 2004.
(d) The most recent version is reference number EN ISO/IEC 17025: 2005, accepted by CEN/CENELEC on 15th March 2005.
(e) ISBN 92 4 154496 1.

2

Guidance

Definition of 'asbestos'

10 'Asbestos' is the general term for the fibrous silicates listed in the definition in regulation 2. Any mixture which contains one or more of these fibrous silicates at more than trace amounts as defined in HSG248 *Asbestos: The analysts' guide for sampling, analysis and clearance procedures (The analyst's guide)*[8] is within the definition. For any work covered by these Regulations 'asbestos' also includes materials containing any of these fibrous silicates or mixtures of these fibrous silicates. Three main types of asbestos were commonly used as follows:

(a) crocidolite (commonly known as blue asbestos);

(b) amosite (commonly known as brown asbestos); and

(c) chrysotile (commonly known as white asbestos).

11 Debris containing asbestos is also covered by these Regulations. If it can be determined that the debris contains raw asbestos, asbestos insulation, asbestos coating or asbestos insulating board then, even though it may not be fulfilling its original purpose, ie the asbestos materials may no longer be coating or insulating anything, a licensed asbestos contractor will be required to carry out the work unless it meets the requirements of regulation 3(2) (see paragraphs 32 and 33).

2

ACOP
2

12 'Asbestos cement' means a material which is predominantly a mixture of cement and chrysotile and which when in a dry state absorbs less than 30% water by weight (see paragraph 16 for details of the test method).

Guidance

13 Asbestos cement is mainly a mixture of chrysotile and cement, which is moulded and compressed to produce a range of asbestos products such as profiled roofing sheets and sidings, flat sheet, gutters, drainpipes, pressure pipes and flues. Asbestos cement was widely used on the exterior of buildings and for drainage products and as it is weatherproof and waterproof will absorb less water (<30%) than asbestos insulation or asbestos insulating board (≥30%). Amosite and/or crocidolite asbestos have also been used in asbestos cement and may sometimes be present along with the chrysotile but in smaller quantities.

14 As the asbestos fibres are mostly firmly bound into the cement and not readily made airborne, work with asbestos cement does not pose the same risks as work with asbestos insulation, asbestos insulating board and sprayed asbestos coatings. Therefore work with asbestos cement will normally fulfil the conditions for regulation 3(2) to apply (see paragraphs 34-39). In such circumstances, work with asbestos cement does not require a licence or notification to the relevant enforcing authority.

15 A competent experienced surveyor will normally be able to visually identify most asbestos cement products. However, if visual identification is inconclusive, then analysis should be carried out to establish the asbestos type(s). If after analysis of the asbestos types there is still doubt about whether a material is asbestos cement product, you will need to carry out a water absorption measurement using the procedure in paragraph 16 to decide whether work on the material requires a licence. Asbestos analysis laboratories should be able to do this test and a list of such laboratories can be obtained from the United Kingdom Accreditation Service (UKAS, 21-47 High Street, Feltham, Middlesex TW13 4UN www.UKAS.com).

2

16 This water absorption test should be carried out on a suitable sample (ie a minimum of 3 cm x 3 cm or 9 cm^2) which is free of any adhering material

Guidance

(partially painted samples can be used but may need longer to absorb water). All asbestos materials should be handled inside a suitable extraction or recirculating air cabinet fitted with high-efficiency filters, or sealed in a suitable container. The following steps should be followed:

(a) Remove the sample of asbestos from packaging/container and either dry for a minimum of 12 hours at 50-110 °C or until the difference between two consecutive weights made at an interval of not less than 1 hour, is less than 1% of the mean of the two measurements.

(b) Before weighing, allow time for the sample to cool and condition.

(c) If weighing outside the containment cabinet, place the sample in a suitable pre-weighed sealable container (eg a sealable plastic bag) and weigh to the nearest 0.01 g. Calculate the dry weight of the sample by subtracting the weight of the container.

(d) After weighing, remove the sample from any container and completely immerse in water for a minimum of 15 minutes and until no more visible signs of bubbles being formed are observed.

(e) If the sample is seen to start to disintegrate during immersion, the test should be terminated and the sample reported as a licensed asbestos material.

(f) If intact, remove the sample from the water and place it on paper towel for 1 minute per side (upper and lower surfaces) to remove any excess surface water.

(g) Place the sample in the pre-weighed sealable container and reweigh. Calculate the wet weight of the sample by subtracting the weight of the container.

(h) Calculate the percentage of water absorbed by the sample using the following equation: ((wet weight - dry weight)/dry weight) x 100.

(i) If the percentage of water absorbed is <30% report as an asbestos material for which a licence is not required to work on (eg asbestos cement). If >30% report as an asbestos material for which a licence is required to work on (eg asbestos insulating board (AIB) or millboard).

2

ACOP

17 Textured decorative coatings which contain asbestos mean thin decorative and textured finishes such as paints and ceiling plasters used to produce visual effects. These coatings are designed to be decorative and any thermal or acoustic properties are incidental to their purpose. The proportion of asbestos in such coatings is normally between 2% and 5% chrysotile.

2

Guidance

18 Work with textured decorative coatings will not normally need to be carried out by a person licensed to work with asbestos, as work with this material will usually fulfil the conditions for regulation 3(2) to apply (see paragraphs 34-39).

19 The term 'coating' does not apply to the base material to which a coating has been applied, even if that base material contains asbestos. (The base material may, however, fall within the definition of asbestos insulation or asbestos insulating board.)

2

ACOP
2

20 'Asbestos insulating board' (AIB) means any flat sheet, tile or building board consisting of a mixture of asbestos and other material except –

ACOP

2

(a) asbestos cement; or

(b) any article of bitumen, plastic, resin or rubber which contains asbestos, and the thermal or acoustic properties of which are incidental to its main purpose.

Guidance

2

21 AIB is a lightly compressed board made from asbestos fibre and hydrated Portland cement or calcium silicate with other filler materials. AIB is covered by this definition whether or not the board is used for insulation. For instance this definition will still apply to asbestos insulating board when its main purpose is structural, eg as a wall partition. Asbestos wallboard (a more compressed variety of AIB) will also fall into this category.

ACOP

2

22 'Asbestos insulation' means any material containing asbestos and used for thermal, acoustic or other insulation purposes (including fire protection) except –

(a) asbestos cement or asbestos insulating board; or

(b) any article of bitumen, plastic, resin or rubber which contains asbestos and the thermal and acoustic properties of which are incidental to its main purpose.

Guidance

2

23 The term 'asbestos insulation' describes asbestos-containing products, which were not in practice applied as coatings: those used for heat, sound, fire protection and other insulation purposes. This includes preformed sections of pipe insulation, asbestos lagging and asbestos infill (asbestos used to fill the spaces between voids, applied between floors and packed around cables where they pass between floors). Millboards are also included in this definition. They have been used for insulation of electrical equipment and for thermal insulation.

ACOP

2

24 'Asbestos coating' means a surface coating which contains asbestos for fire protection purposes or as both heat and sound insulation.

Guidance

2

25 'Asbestos coating' describes the various mixtures containing asbestos, which were widely used as surface coatings for fire protection purposes or as both heat and sound insulation. Most of these coatings were applied by spray but some were applied by hand.

26 'Asbestos coating' does not apply to the base material to which a coating has been applied, even if that base material contains asbestos. (The base material may, however, fall within the definition of asbestos insulation or asbestos insulating board.)

ACOP

2

Work with asbestos

27 'Work with asbestos' includes:

(a) work which consists of the removal, repair or disturbance of asbestos;

(b) work which is ancillary to such work (ancillary work); and

(c) supervising work referred to in sub-paragraphs (a) or (b) above (supervisory work).

28 'Ancillary work' means work associated with the main work of repair, removal or disturbance of asbestos. Work carried out in an ancillary capacity requires a licence unless the main work (ie the removal, repair, disturbance

ACOP

activity) would result in worker exposure which fulfils the conditions for regulation 3(2) to apply. 'Ancillary work' includes the maintenance of equipment which is, or could be, contaminated with asbestos, eg, Class H vacuum cleaners (BS EN 60335[9]) and air extraction equipment (which includes 'negative pressure' units). Paragraph 188 of this ACOP provides more information. 'Negative pressure' refers to air pressure within an enclosure being lower than that outside it. Ancillary work also includes putting up and taking down scaffolding, including any scaffolded frame, to provide access for licensable work where it is foreseeable that the scaffolding activity is likely to disturb the asbestos.

29 'Supervisory work' means work involving direct supervisory control over those removing, repairing or disturbing asbestos. Work carried out in a supervisory capacity requires a licence to work with asbestos unless the work being supervised would result in worker exposure which fulfils the conditions for regulation 3(2) to apply.

Competence

30 Any reference in this ACOP to competence, competent persons or competent employees is a reference to a person or employee who has received adequate information, instruction and training for the task being undertaken and can demonstrate an adequate and up-to-date understanding of the work, required control measures and appropriate law. In addition they must have sufficient experience to apply this knowledge effectively.

2

Regulation 3 — Application of these Regulations

Regulation

(1) These Regulations shall apply to a self-employed person as they apply to an employer and an employee and as if that self-employed person were both an employer and an employee.

(2) Subject to paragraph (3), regulations 8 (licensing), 9 (notification of work with asbestos), 15(1) (arrangements to deal with accidents, incidents and emergencies), 18(1)(a) (asbestos areas) and 22 (health records and medical surveillance) shall not apply where –

(a) the exposure of employees to asbestos is sporadic and of low intensity;

(b) it is clear from the risk assessment that the exposure of any employee to asbestos will not exceed the control limit; and

(c) the work involves –

(i) short, non-continuous maintenance activities,

(ii) removal of materials in which the asbestos fibres are firmly linked in a matrix,

(iii) encapsulation or sealing of asbestos-containing materials which are in good condition, or

(iv) air monitoring and control, and the collection and analysis of samples to ascertain whether a specific material contains asbestos.

(3) No exposure to asbestos will be sporadic and of low intensity within the meaning of paragraph (2)(a) if the concentration of asbestos in the atmosphere

3

Regulation

when measured in accordance with the 1997 WHO recommended method or by a method giving equivalent results to that method approved by the Health and Safety Commission exceeds or is liable to exceed the concentration approved in relation to a specified reference period for the purposes of this paragraph by the Health and Safety Commission.

(4) Where a duty is placed by these Regulations on an employer in respect of his employees, he shall, so far as is reasonably practicable, be under a like duty in respect of any other person, whether at work or not, who may be affected by the work activity carried out by the employer except that the duties of the employer –

(a) under regulation 10 (information, instruction and training) shall not extend to persons who are not his employees unless those persons are on the premises where the work is being carried out; and

(b) under regulation 22 (health records and medical surveillance) shall not extend to persons who are not his employees.

(5) Regulation 17, insofar as it requires an employer to ensure that premises are thoroughly cleaned, shall not apply –

(a) in England and Wales, to a fire and rescue authority within the meaning of section 1 of the Fire and Rescue Services Act 2004,[a] *or in Scotland to a relevant authority within the meaning of section 6 of the Fire (Scotland) Act 2005,*[b] *in respect of premises attended by its employees for the purpose of fighting a fire or in an emergency; or*

(b) to the employer of persons who attend a ship in dock premises for the purpose of fighting a fire or in an emergency, in respect of any ship so attended,

and for the purposes of this paragraph "ship" includes all vessels and hovercraft which operate on water or land and water, and "dock premises" means a dock, wharf, quay, jetty or other place at which ships load or unload goods or embark or disembark passengers, together with neighbouring land or water which is used or occupied, or intended to be used or occupied, for those or incidental activities, and any part of a ship when used for those or incidental activities.

(6) These Regulations shall not apply to the master or crew of a ship or to the employer of such persons in respect of the normal shipboard activities of a ship's crew which are carried out solely by the crew under the direction of the master, and for the purposes of this paragraph "ship" includes every description of vessel used in navigation, other than a ship forming part of Her Majesty's Navy.

3

(a) 2004 c.21; Section 1(2)(d) was amended by the Civil Contingencies Act 2004 (c.36), section 32(1) and Schedule 2 Part 1 paragraph 10(1) and (2).
(b) 2005 asp 5.

Guidance

3

Exceptions from some requirements

31 Where regulation 3(2) applies (ie non-licensable work):

(a) the work will not need to be notified to the relevant Enforcing Authority;

(b) the work will not need to be carried out by holders of a licence to work with asbestos;

(c) the workers will not need to have a current medical and a current health record;

(d) the employer will not need to prepare specific asbestos emergency procedures;

(e) the area around the work does not need to be identified as an asbestos area.

Sporadic and low intensity exposure

32 No exposure to asbestos will be sporadic and of low intensity within the meaning of regulation 3 if the concentration of asbestos in the air exceeds or is liable to exceed 0.6 fibres per cubic centimetre (f/cm^3, which is the same unit as f/ml) in the air measured over a ten-minute period. Work which is likely to result in exposures at or above this level cannot be considered to produce sporadic and low intensity exposure, and therefore the exemptions provided by regulation 3(2) will not apply.

33 When work with the following materials meets the definition of sporadic and low intensity worker exposure then the exemption as provided by regulation 3(2) will apply, but only if it is clear from a suitable and sufficient risk assessment that the control limit of 0.1 f/cm^3 airborne fibres averaged over a 4-hour period will not be exceeded.

ACOP

3

Materials in which the asbestos fibres are firmly linked in a matrix

34 Materials in which the asbestos fibres are firmly linked in a matrix (see regulation 3(2)(c)(ii)) include:

(a) asbestos cement;

(b) textured decorative coatings and paints which contain asbestos;

(c) any article of bitumen, plastic, resin or rubber which contains asbestos where its thermal or acoustic properties are incidental to its main purpose (eg vinyl floor tiles, electric cables, roofing felt).

Guidance

3

35 There may be other materials in which the asbestos fibres can be firmly linked in a matrix such as paper linings, cardboards, felt, textiles, gaskets, washers, and rope where the products have no insulation purposes. If this is the case then the exemption provided in regulation 3(2) may apply.

ACOP

3

36 Therefore, the exceptions listed in paragraph 31 can be applied to most work with these materials but only when a suitable and sufficient risk assessment demonstrates that the control limit will not be exceeded. However the requirements as provided by the remainder of the Regulations will apply to all work with asbestos-containing materials (ACMs). In particular, the work must be undertaken by trained workers in accordance with a plan of work and using proper controls to prevent exposure to and spread of asbestos (see paragraphs 167-176).

37 In general, regulation 3(2) will apply to work with textured decorative coatings containing asbestos and asbestos cement. However the risk assessment may identify factors that lead to the conclusion that the control limit could be exceeded or the exposure would not be sporadic and low intensity and in this case the exemptions would not apply.

Guidance

3

38 Such factors might be a much higher proportion of asbestos in the material than normal, the material being more friable than normal and the best available method of work could result in exposure which could not be considered to be sporadic and low intensity or the control limit being exceeded.

39 HSE has produced guidance on work to which regulation 3(2) will apply including HSG210 *Asbestos essentials task manual: Task guidance sheets for the building maintenance and allied trades.*[10] These task guidance sheets are now available on HSE's website - www.hse.gov.uk/asbestos.

ACOP

3

Short non-continuous maintenance activities

40 The exemption as provided by regulation 3(2) will not apply to work (other than encapsulation and sealing of materials in good condition or air monitoring and control, and the collection and analysis of samples), including supervisory and ancillary activities, with any other types of asbestos materials unless it is a short, non-continuous maintenance activity (regulation 3(2)(c)(i)) and a suitable and sufficient risk assessment demonstrates that the control limit will not be exceeded and that any exposure will be sporadic and low intensity. The type of work that meets the exemption criteria for asbestos insulation and asbestos insulation board is set out in paragraph 41.

Asbestos insulation and insulating board

41 Due to the relative ease with which asbestos fibres can be released when working with asbestos insulation and insulating board, in most circumstances work with these materials should only be carried out by those holding a licence. The exemption from licensing only applies to short duration work where the risk assessment shows that the work will only produce sporadic and low intensity exposure as defined, and will not exceed the control limit. The work can only be considered as short duration if any one person carries out work with these materials for less than one hour in a seven-day period. The total time spent by all workers on the work should not exceed a total of two hours. When calculating the time the work takes you should include anything ancillary to the work which is liable to disturb the asbestos, including setting up enclosures and clearing any potentially affected area.

Guidance

3

42 Examples of short, non-continuous maintenance activities when working with asbestos insulating board are included in the HSE publication *Asbestos essentials task manual*[10] (HSG210) and on HSE's asbestos website - www.hse.gov.uk/asbestos. HSE is considering providing additional guidance on encapsulation and analysis as detailed in regulation 3(2)(c)(iii) and (iv).

Employers' duties

43 These Regulations place specific duties on employers, self-employed people and employees (eg see paragraphs 203-204). Table 1 summarises the scope of the employer's (and self-employed people's) duties in respect of employees and other people.

ACOP

3

44 Employers must take into account people other than their own employees in the assessment required by regulation 6 and in the action taken to prevent or control exposure required by regulation 11.

ACOP

3

45 Whenever two or more employers work with asbestos or are likely to come into contact with asbestos at the same time at the same workplace they should co-operate in order to meet their separate responsibilities towards their own and each other's employees as well as other people who may be affected by the work, and should also consult relevant safety representatives.

Guidance

Table 1 Summary of employers' (and self-employed people's) duties in respect of employees and others

Duty of employer relating to:		*Duty for the protection of:*	
	Employees	*Other people on the premises*	*Other people likely to be affected by work*
Regulations 5-9, 11, 13-15, 17-19 and 23	Yes	SFAIRP	SFAIRP
Regulation 10 – provision of information, instruction and training	Yes	SFAIRP	No
Regulation 22 – health records and medical surveillance	Yes	No	No

Note: SFAIRP means 'so far as is reasonably practicable'

Duties under other Regulations

46 There are other people associated with the work covered by this ACOP who may not have direct duties under the Asbestos Regulations but may well have duties under other legislation. This may include analysts, clients (apart from regulation 4 of the Asbestos Regulations), planning supervisors, designers and principal contractors as defined by the Construction (Design and Management) Regulations (CDM) as amended.[7] People carrying out site clearance certification (eg analysts) may have general duties under sections 3, 7, 8, 36 or 37 of the the HSW Act.

3

PART 2 GENERAL REQUIREMENTS

Regulation 4

Duty to manage asbestos in non-domestic premises

Regulation

(1) In this regulation "the dutyholder" means –

(a) every person who has, by virtue of a contract or tenancy, an obligation of any extent in relation to the maintenance or repair of non-domestic premises or any means of access thereto or egress therefrom; or

(b) in relation to any part of non-domestic premises where there is no such contract or tenancy, every person who has, to any extent, control of that part of those non-domestic premises or any means of access thereto or egress therefrom,

and where there is more than one such dutyholder, the relative contribution to be made by each such person in complying with the requirements of this regulation will be determined by the nature and extent of the maintenance and repair obligation owed by that person.

(2) Every person shall cooperate with the dutyholder so far as is necessary to enable the dutyholder to comply with his duties under this regulation.

(3) In order to enable him to manage the risk from asbestos in non-domestic premises, the dutyholder shall ensure that a suitable and sufficient assessment is carried out as to whether asbestos is or is liable to be present in the premises.

(4) In making the assessment –

(a) such steps as are reasonable in the circumstances shall be taken; and

(b) the condition of any asbestos which is, or has been assumed to be, present in the premises shall be considered.

(5) Without prejudice to the generality of paragraph (4), the dutyholder shall ensure that –

(a) account is taken of building plans or other relevant information and of the age of the premises; and

(b) an inspection is made of those parts of the premises which are reasonably accessible.

(6) The dutyholder shall ensure that the assessment is reviewed forthwith if –

(a) there is reason to suspect that the assessment is no longer valid; or

(b) there has been a significant change in the premises to which the assessment relates.

4 *(7) The dutyholder shall ensure that the conclusions of the assessment and every review are recorded.*

Regulation

(8) Where the assessment shows that asbestos is or is liable to be present in any part of the premises the dutyholder shall ensure that –

(a) a determination of the risk from that asbestos is made;

(b) a written plan identifying those parts of the premises concerned is prepared; and

(c) the measures which are to be taken for managing the risk are specified in the written plan.

(9) The measures to be specified in the plan for managing the risk shall include adequate measures for –

(a) monitoring the condition of any asbestos or any substance containing or suspected of containing asbestos;

(b) ensuring any asbestos or any such substance is properly maintained or where necessary safely removed; and

(c) ensuring that information about the location and condition of any asbestos or any such substance is –

(i) provided to every person liable to disturb it, and

(ii) made available to the emergency services.

(10) The dutyholder shall ensure that –

(a) the plan is reviewed and revised at regular intervals, and forthwith if –

(i) there is reason to suspect that the plan is no longer valid, or

(ii) there has been a significant change in the premises to which the plan relates;

(b) the measures specified in the plan are implemented; and

(c) the measures taken to implement the plan are recorded.

(11) In this regulation, a reference to –

(a) "the assessment" is a reference to the assessment required by paragraph (3);

(b) "the premises" is a reference to the non-domestic premises referred to in paragraph (1); and

4 *(c) "the plan" is a reference to the plan required by paragraph (8).*

Guidance

47 Owners and occupiers of non-domestic premises, who have maintenance and repair responsibilities for those premises, have a duty to assess them for the presence of asbestos and the condition of that asbestos. Where asbestos is present the dutyholder must ensure that the risk from the asbestos is assessed, that a written plan identifying where that asbestos is located is prepared and that measures to manage the risk from the asbestos are set out in that plan and are implemented. Other parties have a legal duty to co-operate with the dutyholder.

4

Guidance
4

48 This ACOP does not deal with regulation 4 which has its own ACOP entitled, *The management of asbestos in non-domestic premises.*

Regulation 5

Identification of the presence of asbestos

Regulation

An employer shall not undertake work in demolition, maintenance, or any other work which exposes or is liable to expose his employees to asbestos in respect of any premises unless either –

(a) he has carried out a suitable and sufficient assessment as to whether asbestos, what type of asbestos, contained in what material and in what condition is present or is liable to be present in those premises; or

(b) if there is doubt as to whether asbestos is present in those premises he –

(i) assumes that asbestos is present, and that it is not chrysotile alone, and

(ii) observes the applicable provisions of these Regulations.

5

Guidance

Identification of asbestos

49 As part of the management plan required by regulation 4, occupiers or owners of premises have an obligation to inform any person liable to disturb ACMs, including maintenance workers, about the presence and condition of such materials.

50 If work to be carried out is part of a larger project which attracts the requirements of the Construction (Design and Management) Regulations (CDM) 1994, as amended, the health and safety plan prepared by the planning supervisor should contain information on whether the materials contain asbestos and what type they are.

51 The employer should not simply rely on the information provided by other dutyholders unless the dutyholder can produce reasonable evidence to confirm the validity of the information such as survey details (eg a type 3 survey is required for major refurbishment and removal work covered by the CDM Regulations - see MDHS100 *Surveying, sampling and assessment of asbestos-containing materials*[11]) and provide information on the nature of suspect material (eg the analytical report and/or management plan for the area of work should be made available).

5

ACOP

52 If appropriate information for the scope of work to be undertaken is not available or is not in a reliable form, then before carrying out any work involving the potential disturbance of the fabric of a building the employer should either:

(a) establish whether the part of the building that is likely to be disturbed contains asbestos, and if so the type. This may require a survey and analysis of representative samples; or

(b) assume that the part of the building being worked upon contains the most hazardous types of asbestos, crocidolite (commonly known as blue asbestos) or amosite (commonly known as brown asbestos) and take the precautions outlined in the Regulations and this ACOP for licensable work.

5

Regulation 6

Assessment of work which exposes employees to asbestos

Regulation

(1) An employer shall not carry out work which is liable to expose his employees to asbestos unless he has –

(a) made a suitable and sufficient assessment of the risk created by that exposure to the health of those employees and of the steps that need to be taken to meet the requirements of these Regulations;

(b) recorded the significant findings of that risk assessment as soon as is practicable after the risk assessment is made; and

(c) implemented the steps referred to in sub-paragraph (a).

(2) Without prejudice to the generality of paragraph (1), the risk assessment shall –

(a) subject to regulation 5, identify the type of asbestos to which employees are liable to be exposed;

(b) determine the nature and degree of exposure which may occur in the course of the work;

(c) consider the effects of control measures which have been or will be taken in accordance with regulation 11;

(d) consider the results of monitoring of exposure in accordance with regulation 19;

(e) set out the steps to be taken to prevent that exposure or reduce it to the lowest level reasonably practicable;

(f) consider the results of any medical surveillance that is relevant; and

(g) include such additional information as the employer may need in order to complete the risk assessment.

(3) The risk assessment shall be reviewed regularly, and forthwith if –

(a) there is reason to suspect that the existing risk assessment is no longer valid;

(b) there is a significant change in the work to which the risk assessment relates; or

(c) the results of any monitoring carried out pursuant to regulation 19 show it to be necessary,

and where, as a result of the review, changes to the risk assessment are required, those changes shall be made and, where they relate to the significant findings of the risk assessment or are themselves significant, recorded.

(4) Where, in accordance with the requirement in paragraph (2)(b), the risk assessment has determined that the exposure of his employees to asbestos may exceed the control limit, the employer shall keep a copy of the significant findings of the risk assessment at those premises at which, and for such time as, the work to which that risk assessment relates is being carried out.

6

Guidance
6

General requirements for risk assessments

53 Regulation 11(1) places a duty on employers to prevent the exposure of his employees to asbestos so far as is reasonably practicable and this should be the first consideration. Employers must first decide whether it is possible to avoid exposure to asbestos altogether. The risk assessment must consider how this can be achieved and also any risks which complying with this duty may present. For example, if cables are rerouted to avoid disturbing ACMs, the risk assessment should consider what other risks the workers would face and what the overall risk would be.

ACOP
6

54 If work which is liable to expose employees to asbestos is unavoidable, then before starting the work, employers must make a suitable and sufficient assessment of the risks created by the likely exposure to asbestos of employees and others who may be affected by the work and identify the steps required to be taken by the Asbestos Regulations.

Guidance
6

55 The purpose of the assessment is to ensure that people properly consider the scope of the proposed works to establish the extent of the potential risks so that they can identify which legal provisions apply (including whether the work is licensable) and to determine the most appropriate work methods to comply with legal duties. This should help ensure that exposure to asbestos is prevented or adequately controlled so that the health of employees and other people is not put at risk, and that other risks are also properly controlled.

56 The employer should ensure that there is effective communication between the various parties (such as the client, Principal Contractor and other employers) as is necessary on any site to minimise the risk of exposure to asbestos and to minimise the overall risks.

ACOP
6

57 The plan of work should always be job-specific. However, information from previous similar jobs can be used in the risk assessment provided there are no additional risks and it is appropriate for the site conditions.

58 Where the degree and nature of the work varies significantly from site to site, for example in demolition or refurbishment, or where the type of asbestos material varies, a new, site-specific assessment and plan of work (see regulation 7) will be necessary.

59 The assessment should be done in sufficient time to ensure compliance with the requirements of the Regulations and to enable the appropriate precautions to be undertaken before work commences. The significant findings of the assessment should be in writing, and should form the basis of the plan of work (regulation 7).

Guidance
6

60 Employers may also have duties under other legislation to carry out a separate risk assessment. For example, if employees are likely to be exposed to other risks such as falls from height, confined spaces or hot conditions, assessments will be required under the Management of Health and Safety at Work Regulations 1999 (MHSW), as amended.[12]

61 For licensable work, a copy of the risk assessment relating to the work must be kept at the premises where the work is being undertaken.

ACOP
6

62 The significant findings of risk assessment should be communicated in a comprehensible way to the employees before they start the work and others as appropriate to minimise the risks to them.

Guidance
6

63 Employers have duties under the Health and Safety (Consultation with Employees) Regulations[5] to consult their employees. The employees should be consulted on the risk assessment to help determine the nature and degree of their exposure. Proper consultation can also make a significant contribution to creating and maintaining a safe and healthy working environment and an effective health and safety culture.

ACOP
6

64 To decide whether or not the control limit is likely to be exceeded, it is first necessary to know what concentration of asbestos fibres are likely to be present in the air. It will be necessary to confirm the estimated exposures by measurement, using an appropriate method (see paragraphs 328-338) unless there is already sufficient, relevant and reliable data available. A summary of this data and the source from which it was derived should be included in the assessment.

Guidance
6

65 Guidance on methods approved by HSC may be found in *The analysts' guide*.[8]

ACOP
6

66 Employers must ensure that whoever carries out the assessment and provides advice on the prevention and control of exposure is competent to do so in accordance with regulation 10. This does not necessarily mean that particular qualifications are required. However, whoever carries out the assessment should:

(a) have adequate knowledge, training and expertise in understanding the risks from asbestos and be able to make informed and appropriate decisions about the risks and precautions that are needed;

(b) know how the work activity may disturb asbestos;

(c) be familiar with and understand the requirements of the Asbestos Regulations and this ACOP;

(d) have the ability and the authority to collate all the necessary, relevant information; and

(e) be able to assess other non-asbestos risks on site.

67 To be suitable and sufficient, the risk assessment should include:

(a) for non-licensed work, a statement of the reasons why the work with asbestos will fulfil the conditions for regulation 3(2) to apply and will not therefore be work which requires a licence;

(b) a description of the work (eg repair, removal, encapsulation of ACM, maintenance and testing of plant and equipment contaminated with ACMs), and the expected scale and duration;

(c) a description of the type(s) of asbestos present and the results of any analysis or a statement that the asbestos is not chrysotile alone;

(d) a description of the quantity, form, size, means of attachment, extent and condition of any ACMs present;

(e) details of expected exposures, noting:

(i) whether they are liable to exceed the control limit and the number of people likely to be affected;

ACOP

(ii) the level of the expected exposure, so that suitable respiratory protective equipment (RPE) can be assessed and selected;

(iii) whether anyone other than employees may be exposed, and their expected exposures;

(iv) whether intermittent higher exposures may arise and their expected frequency and duration; and

(v) results already available from air monitoring in similar circumstances;

(f) the steps to be taken to control exposure to the lowest level reasonably practicable, for example for licensable work, the type of controlled wetting and method of application, the use of local exhaust ventilation (LEV) (eg shadow vacuuming), glovebag and wrap and cut and for non-licensable building work, the use of low dust methods, shadow vacuuming, wetting etc (see paragraphs 155-189);

(g) the steps taken to control the release of asbestos into the environment, for example enclosures with negative pressure and entry and exit procedures. Where it is not considered practicable to use an enclosure, a full justification is required, and what action should be taken if an accidental release was to occur;

(h) details of the decontamination procedures including the use of hygiene units where appropriate;

(i) procedures for the selection, provision, use and decontamination of personal protective equipment (PPE) which includes respiratory protective equipment (RPE);

(j) procedures for the removal of waste and contaminated tools and equipment from the work area and the site;

(k) procedures for dealing with emergencies, including, for example, those associated with working in confined spaces;

(l) any other information relevant to safe working practices such as other significant non-asbestos hazards like working at heights or in confined spaces; and

(m) management arrangements ensuring that risks are adequately controlled during the work.

6 68 The findings of the assessment detailed in paragraph 67 are all deemed to be 'significant' and must be recorded as required by regulation 6(1)(b).

Guidance

6 69 Knowing the type of ACM (eg AIB, asbestos insulation, asbestos coating, asbestos cement, asbestos-containing textured decorative coatings) is necessary to estimate the potential fibre release for assessment purposes; to select the most appropriate handling and removal techniques, as appropriate or combinations of techniques; and to determine whether the work will be licensable. It is essential, for example, to be able to distinguish between asbestos cement and AIB. Where there is doubt employers should err on the side of caution and assume the material is insulating board and take precautions accordingly. For ancillary work involving the testing and maintenance of plant and equipment the asbestos is most likely to be in the form of dust and the type of ACM may not be relevant. The condition of the material can have a significant effect on the assessment. Knowing the extent of

Guidance 6

the material (eg its length and span, whether it extends into other rooms and work areas) is also important so that the number of enclosures required, and the necessary arrangements for the transfer of waste, can be properly assessed. This will avoid any confusion over what work is being done and which ACMs will remain in place, or detail of the areas where a certificate of reoccupation will be sought.

ACOP 6

Further risk assessment requirements for licensable work

70 For licensable work, to be suitable and sufficient the assessment should, in addition to the elements in paragraph 67, also include:

(a) the reasons for the chosen work method. Except under exceptional circumstances it is not justifiable to work with licensable materials when the material is dry or the environment hot or with the use of power tools which disturb asbestos fibres (see paragraphs 72-73).

(b) the arrangements required to ensure that the premises or parts of premises where the work has taken place are left clean and safe for reoccupation. These should include:

- **(i) detail of the areas where a certificate of reoccupation will be sought;**
- **(ii) consideration of potential problems for a certificate of reoccupation, eg earth floors, limpet spray ingrained in concrete or tar-like layers, wet areas which cannot be dried out and the presence of ACMs which are intended to remain in the areas after the work is complete;**
- **(iii) consideration of the need for pre-cleaning (often required before the setting up of any enclosure) including the control measures required to prevent the release of asbestos fibres.**

71 The elements listed in paragraph 70 are all deemed to be 'significant' and must be recorded as required by regulation 6(1)(b).

72 Hot work with asbestos is to be avoided. Hot work will only be permissible in rare and exceptional circumstances when all possible alternatives have been considered. If it can be justified then additional precautions will need to be taken to prevent heat stress and other risks. This will include reduced work periods and the effective introduction of cool air to the work area (eg air conditioning).

Guidance 6

73 Asbestos removal and hot working are a difficult combination to manage and control effectively. In particular, the various precautions necessary to protect workers from exposure to asbestos dust and to prevent its spread can result in a greatly increased thermal health risk. When operators have left the work area and have decontaminated then they may need to be provided with suitable drinks etc to prevent dehydration. In addition to the heat stress issues, hot work can also lead to deterioration in asbestos controls. All avenues should be explored to remove the heat source including the provision of alternative plant. Wherever possible hot plant should be shut down or turned off and allowed to cool before asbestos removal work commences. This may mean scheduling work to be done during plant shutdown or annual holiday. When hot conditions are due to the climate then work should be scheduled in the evening or overnight. Where work arises at short notice through incidents or emergencies, then short-term remedial action should be taken as far as possible (eg by making temporary repairs or encapsulation) until the work can be incorporated into a programmed plant shut down and carried out with the plant cold.

ACOP

Reviewing assessments

74 Employers should review risk assessments and, as appropriate, plans of work as part of the ongoing management of their health and safety systems and to ensure that the principles of good practice have been applied. The review should be conducted by a competent person. A specific review should also take place if:

(a) fibre control methods change;

(b) there is doubt about the efficiency of control measures;

(c) there is a significant change in the type of work, amount of asbestos or method of work; or

(d) the results of any air monitoring indicate the exposure levels to be higher than previously assessed.

75 Where monitoring of exposure levels, or other information gathered during the course of work, indicates that the initial assessment was wrong in respect of either the duration of the task or the nature of the materials concerned, consideration should be given to reviewing the assessment and control measures and indeed whether the nature and extent of the exposure means that the work should be undertaken using different methods and equipment. Where work has been deemed not to require a licensed contractor that decision may need to be reviewed. Any changes subsequently made to the assessment and hence plans of work (regulation 7) must be recorded in writing.

6

Regulation 7

Plans of work

Regulation

(1) An employer shall not undertake any work with asbestos unless he has prepared a suitable written plan of work detailing how that work is to be carried out.

(2) The employer shall keep a copy of the plan of work at those premises at which the work to which the plan relates is being carried out for such time as that work continues.

(3) In cases of final demolition or major refurbishment of premises, the plan of work shall, so far as is reasonably practicable, and unless it would cause a greater risk to employees than if the asbestos had been left in place, specify that asbestos shall be removed before any other major works begin.

(4) The plan of work shall include in particular details of –

(a) the nature and probable duration of the work;

(b) the location of the place where the work is to be carried out;

(c) the methods to be applied where the work involves the handling of asbestos or materials containing asbestos;

(d) the characteristics of the equipment to be used for –

(i) protection and decontamination of those carrying out the work, and

7

Regulation

(ii) protection of other persons on or near the worksite;

(e) the measures which the employer intends to take in order to comply with the requirements of regulation 11; and

(f) the measures which the employer intends to take in order to comply with the requirements of regulation 17.

(5) The employer shall ensure, so far as is reasonably practicable, that the work to which the plan of work relates is carried out in accordance with that plan and any subsequent written changes to it.

7

ACOP

Plan of work

76 For any work involving asbestos, including maintenance work that may disturb it, the employer of the workers involved must draw up a written plan of how the work is to be carried out before work starts. Employers must make sure their employees follow the plan of work (sometimes called a method statement) so far as it is reasonably practicable to do so. Where unacceptable risks to health and/or safety are discovered while work is in progress, for example disturbance of hidden, missed or incorrectly identified ACMs, any work affecting the asbestos should be stopped except for that necessary to render suitable control and prevent further spread (see paragraphs 247-251 for further guidance). Where there is extensive damage to ACMs which causes contamination of the premises, or part of the premises, then the area should be immediately evacuated. Work should not restart until a new plan of work is drawn up or until the existing plan is amended. Some measures, for example, should only be carried out by licensed contractors.

77 For licensable work in particular, the plan of work should identify procedures to adopt in emergencies and indicate clearly what remedial measures can be undertaken by staff.

7

Guidance

78 In the case of final demolition or major refurbishment, the plan of work must specify that all asbestos is removed before any other major work begins where this is reasonably practicable and does not cause a greater risk to employees than if the asbestos had been left in place. Where removal of ACMs is time-consuming and resource-intensive and only involves a lower risk material such as textured decorative coatings containing asbestos, then removal prior to demolition or major refurbishment may not be reasonably practicable.

79 The plan of work must include the following information:

(a) the nature and probable duration of the work;

(b) the number of persons involved in the work;

(c) the address and location where the work is to be carried out;

(d) the methods to be used to prevent or reduce exposure to asbestos, for example, the prevention and control measures, the arrangements for keeping premises and plant clean and the arrangements for the handling and disposing of asbestos waste;

(e) the type of equipment, including PPE, used for:

(i) the protection and decontamination of those carrying out the work; and

7

Guidance
7

(ii) the protection of other people present at or near the worksite.

80 It will usually be necessary for the plan to include the site layout and a description of the location and nature of the asbestos present and which ACMs will be disturbed by the work.

ACOP
7

81 Work should not take place unless a copy of the plan of work is readily available on site. Employees should be told what the plan contains and instructed on the work methods and controls to be used. The plan of work should also be brought to the attention of anyone who needs to see it, including those carrying out the visual inspection and/or air clearance monitoring once the work or section of work has come to an end. Employers should make a copy of the plan of work available on request to employees, safety representatives and other elected representatives of employee health and safety, as well as others who may be affected by the work.

Guidance
7

82 Arrangements should be made to ensure that work is carried out in accordance with the plan of work.

ACOP
7

Further information to be included for licensable work

83 In addition to the information specified above, when licensable work is being carried out, the plan of work should be site specific and cover in sufficient detail the following information:

(a) the scope of the work as identified by the risk assessment;

(b) details of the hygiene facilities, transit route and decontamination arrangements, vacuum cleaners, other equipment, air monitoring, protective clothing and RPE, communication between the inside and outside of the enclosure; and

(c) details of the use of barriers and signs, location of enclosures and airlocks, location of skips, negative pressure units, air monitoring, cleaning and clearance certification, emergency procedures.

Guidance
7

84 As good practice, other items could be included in the plan such as details of checks undertaken for other hazards, the name(s) of the supervisor(s), the name of organisation that will undertake site clearance certification (see paragraphs 339-342), and details of any nearby ACMs and their extent so that there is no confusion over what work is being done and which ACMs will remain in place.

85 Generic assessments may form a useful starting point for plans of work/method statements but they need to be developed into documents that identify and address site-specific issues.

86 For the majority of licensed contractors it is a condition of their licence to notify the appropriate enforcing authority 14 days in advance of each job with specified information which is also contained in the plan of work.

Regulation 8

Licensing of work with asbestos

Regulation

(1) Subject to regulation 3(2), an employer shall not undertake any work with asbestos unless he holds a licence granted under paragraph (2) of this regulation.

(2) The Executive may grant a licence for work with asbestos if it considers it appropriate to do so and –

(a) the person who wishes the licence to be granted to him has made application for it on a form approved for the purposes of this regulation by the Executive; and

(b) the application was made at least 28 days before the date from which the licence is to run, or such shorter period as the Executive may allow.

(3) A licence under this regulation –

(a) shall come into operation on the date specified in the licence, and shall be valid for any period up to a maximum of three years that the Executive may specify in it; and

(b) may be granted subject to such conditions as the Executive may consider appropriate.

(4) The Executive may vary the terms of a licence under this regulation if it considers it appropriate to do so and in particular may –

(a) add further conditions and vary or omit existing ones; and

(b) reduce the period for which the licence is valid or extend that period up to a maximum of three years from the date on which the licence first came into operation.

(5) The Executive may revoke a licence under this regulation if it considers it appropriate to do so.

(6) The holder of a licence under this regulation shall return the licence to the Executive –

(a) when required by the Executive for any amendment; or

8 *(b) following its revocation.*

ACOP

Licensing of work with asbestos

87 This regulation means that you must not carry out any work with asbestos (other than that fulfilling the conditions for regulation 3(2)) unless you hold a licence issued under this regulation and comply with its terms and conditions. This includes:

(a) supervisory work;

(b) ancillary work;

8 (c) work with asbestos in your own premises with your own employees; and

ACOP

(d) supply of labour.

88 Work within the scope of licensing includes work with asbestos insulation, asbestos coatings (excluding most work with textured decorative coatings containing asbestos) and asbestos insulating board.

89 For supervisory work you need a licence when directly supervising licensable work but not if you are:

(a) the client who has engaged a licensed contractor to do the licensable work;

(b) the principal or main contractor on a construction or demolition site if the licensable work is being done by a subcontractor holding an asbestos licence;

(c) an analyst checking that the area is clear of asbestos at the end of a job;

(d) carrying out quality control work such as:

  (i) atmospheric monitoring outside enclosures while asbestos removal work is in progress; or

  (ii) checking outside enclosures that work has been carried out to a standard which meets the terms of the contract;

(e) a consultant or other reviewing tender submissions on behalf of the client.

90 For ancillary work, you will need a licence for:

(a) setting up and taking down enclosures for notifiable and licensed asbestos work;

(b) putting up and taking down scaffolding, including any scaffolded frame, to provide access for licensable work where it is foreseeable that the scaffolding activity is likely to disturb the asbestos;

(c) maintaining air extraction equipment (which includes 'negative pressure' units);

(d) work done within an asbestos enclosure, such as sealing an electric motor in polythene and installing ducting to the motor to provide cooling air from outside the enclosure; and

(e) cleaning the structure, plant and equipment inside the enclosure.

91 As a licence holder, you will be required:

(a) to notify the work to the appropriate enforcing authority (regulation 9);

(b) to ensure medical surveillance is carried out for your employees/yourself (regulation 22);

(c) to maintain health records for employees/yourself (regulation 22);

(d) to prepare procedures in case of emergencies (regulation 15(1)); and

8

(e) demarcate the work areas appropriately (regulation 18(1)(a)).

92 All licences issued for work with asbestos are granted by HSE under the terms of this regulation. Fees are payable for issuing licences, reassessments and changes to licences.

Applications for licences

93 You need to make applications for licences and for the renewal of licences on the approved form (FOD ASB1). The form is available from:

The Health and Safety Executive
Asbestos Licensing Unit
Belford House
59 Belford Road
Edinburgh, EH4 3UE

Tel: 0131 247 2135

94 The Regulations allow you to apply for a licence to do work with asbestos. Before the licence can be granted you will have to:

(a) show adequate knowledge of the Health and Safety at Work etc Act 1974, the Control of Asbestos Regulations 2006, this Approved Code of Practice, and other guidance on work with asbestos materials;

(b) demonstrate competence for the types of work you intend to carry out with asbestos;

(c) intend to carry out work for which a licence is required within the licence period.

95 However, you may wish to work with only one type of material, eg asbestos insulation. In this case, the licence, if issued, may only allow you to work with this material alone.

96 Your application will need to reach the Asbestos Licensing Unit at least 28 days before the date from which you wish the licence to run. In some circumstances, the HSE may be prepared to issue a licence in a shorter period. Renewal applicants should note that it is not possible to extend an existing licence beyond its expiry date.

ACOP
8

97 The person(s) signing the application form will be required to declare that:

(a) the information provided by the Asbestos Licensing Unit has been read and understood by (one of) the signatory, directors, partners, person responsible within the organisation for asbestos operations;

(b) they have the appropriate authority within the organisation to represent and bind the company;

(c) the organisation intends to carry out work with ACMs for which a licence is required under the Control of Asbestos Regulations 2006;

(d) to the best of their knowledge the answers given in the application are correct;

ACOP

8

(e) **they understand it is an offence to make a false declaration, which may result in an asbestos licence being revoked and/or prosecution;**

(f) **they have informed their employees of the application.**

Guidance

98 HSE issues all licences, even if licence holders carry out all their work with asbestos within premises that are inspected by local authority (LA) inspectors (these premises are listed in Table 2). Only HSE can amend the terms or conditions of, or revoke, a licence. HSE's Asbestos Licensing Unit works closely with inspectors in all the relevant enforcing authorities to keep records of the activities and performance of licence holders and to consider whether any changes need to be made to the conditions imposed on a licence holder.

The licence

99 The licence will specify the terms and conditions laid down by HSE. The conditions imposed on you will depend upon HSE's assessment of your application. This includes a check on the information you give in the application form as well as an examination of your current performance record if you already hold a licence. HSE may refuse to issue a licence to you:

(a) if you have been convicted of a criminal offence involving work with asbestos;

(b) where a pattern of poor performance has emerged over several site visits, demonstrating evidence of poor working conditions and control. This may have resulted in enforcement action (eg conviction(s) for asbestos-related offences, enforcement notices for asbestos-related deficiencies, warning letters etc);

(c) there has been an extremely serious incident where significant breach(es) of asbestos-related legislation have occurred. The failures that led to the breaches may be so significant that it is considered necessary to initiate revocation proceedings irrespective of whether or not enforcement action has occurred;

(d) if you cannot demonstrate that you have adequate knowledge or arrangements in place to protect the health of your employees and others during work with the relevant ACMs;

(e) if you have been found guilty of health and safety offences;

(f) if you have had two enforcement notices issued against you (related to any health and safety matter) within a two-year period;

(g) if you have previously failed to comply with the conditions and limitations of a licence to work with asbestos;

(h) where Directors or Senior Managers have had a significant involvement in circumstances that have been considered suitable for licence refusal or revocation; or

8

(i) if you have breached legislation, even if not enforced by HSE, LAs or other relevant enforcing authority, which brings into doubt your reputation to be a licence holder.

Guidance 8

Period of issue

100 All licences are issued for a limited period of time so that HSE can regularly review your performance. New applicants are issued with an initial licence for a shorter period (usually for one year). When this is put forward for renewal the period is generally for three years unless you have a record of poor performance or you have not undertaken any work under the terms of your licence. In this case the period may be reduced to a period deemed appropriate by HSE, or the licence may not be renewed. The maximum period for a licence is three years.

101 If you have little previous experience of work with asbestos, or the relevant enforcing authority (see Table 2) has not inspected your work, the terms of the licence may be more restrictive than if you are an experienced contractor who is well known to the relevant enforcing authority.

102 HSE may amend a licence to impose a shorter licence period on you if they want to check that you are taking adequate health and safety precautions during your work or there is doubt of your continuing competence to work with asbestos. This includes work with asbestos and any other work you carry out where you have failed to take adequate health and safety precautions. You may be informed by HSE that they may refuse to renew the licence when it is due to end if you do not improve your standard of protection for employees.

Conditions that may be included in the licence form

103 Regulation 8(4)(a) allows the HSE to impose whatever conditions on the licence it considers appropriate.

ACOP 8

104 As a condition of your licence, you will normally also be required to submit additional documentation as part of the notification process, including a copy of your plan of work.

105 In the case of particularly difficult jobs, for example where the work may need extra precautions to those set out in published guidance, the HSE may impose special conditions on you. These should be reflected in the plan of work (method statement) you send to the enforcing authority.

Guidance 8

106 HSE may also impose other conditions on you if there are concerns about your ability to work in accordance with the appropriate standards or if the work carries particular risks.

107 However, HSE will use the conditions of licensing to monitor closely the work of certain licence holders without placing unnecessary limitations or conditions on the work of competent employers and self-employed contractors.

Revocation of licences

108 HSE has the power, under this regulation, to revoke licences where it considers it appropriate to do so. HSE will consider revocation if, for example:

(a) if you have been convicted of a criminal offence involving work with asbestos;

(b) where a pattern of poor performance has emerged over several site visits, demonstrating evidence of poor working conditions and control. This may have resulted in enforcement action (eg conviction(s) for asbestos-related

Guidance

8

offences, enforcement notices for asbestos-related deficiencies, warning letters etc);

(c) there has been an extremely serious incident where significant breach(es) of asbestos-related legislation have occurred. The failures that led to the breaches may be so significant that it is considered necessary to initiate revocation proceedings irrespective of whether or not enforcement action has occurred;

(d) if you cannot demonstrate that you have adequate knowledge or arrangements in place to protect the health of your employees and others during work with the relevant ACMs;

(e) if you have been found guilty of health and safety offences;

(f) if you have had two enforcement notices issued against you within a two-year period;

(g) if you have previously failed to comply with the conditions and limitations of a licence to work with asbestos;

(h) where Directors or Senior Managers have had a significant involvement in circumstances that have been considered suitable for licence refusal or revocation; or

(i) if you have breached legislation, even if not enforced by HSE, LAs or other relevant enforcing authority, which brings into doubt your reputation to be a licence holder.

109 Revocation is a very serious step, which could affect the livelihood of you and your employees. HSE therefore considers each case (and may, depending on the circumstances, go through a number of stages such as warning letters, enforcement notices, formal interviews, review board meetings etc) before revoking a licence. HSE will take into consideration the performance history of your company, any available evidence, and the particular circumstances before deciding whether or not to revoke your licence.

110 HSE's revocation policy is available on the asbestos section of the HSE website: www.hse.gov.uk/asbestos.

Appeals

111 You can make informal representations, in writing, against a decision to either revoke or not to renew your licence to the senior HSE manager directly responsible for the Asbestos Licensing Unit at the address in paragraph 93. If the matter is still not resolved, you can appeal to the Secretary of State for Work and Pensions under section 44 of the Health and Safety at Work etc Act 1974. Normally the Secretary of State will appoint a person with relevant legal and/or practical experience to hear the appeal. The appeal may be decided on the basis of written submissions, but if either you or HSE want to be heard, that opportunity will be given.

112 If you appeal against a decision you should include:

(a) your name and address;

(b) a photocopy of your current licence;

Guidance 8

(c) the grounds for your appeal.

113 You should address your appeal to:

The Secretary of State for Work and Pensions
Department for Work and Pensions (DWP)
79 Whitehall
London SW1A 2NS

114 You can get more guidance on the appeals system from the enforcing authority.

Penalties

115 If you are convicted in the Crown Court/High Court for carrying out work for which you do not hold a licence or for breaching a term or condition attached to your licence, you can be fined an unlimited amount, imprisoned for a term of up to two years, or both.

116 If the case is heard in a Magistrates or Sheriff Court the maximum penalty is £5000 per convicted offence.

Regulation 9

Notification of work with asbestos

Regulation 9

(1) Subject to regulation 3(2), an employer shall not undertake any work with asbestos unless he has notified the appropriate office of the enforcing authority in writing of the particulars specified in Schedule 1 at least 14 days before commencing that work or such shorter time before as the enforcing authority may agree.

(2) Where an employer has notified work in accordance with paragraph (1) and there is a material change in that work which might affect the particulars so notified (including the cessation of the work), the employer shall forthwith notify the appropriate office of the enforcing authority in writing of that change.

Guidance 9

Notification of licensable work

117 If you undertake licensable work you have to notify the appropriate enforcing authority with details of the proposed work. This gives the enforcing Authority the opportunity to assess your proposals for carrying out work with asbestos and to inspect the site either before or during the work.

118 You will normally be required to notify the relevant enforcing authority office 14 days before any work begins, but the enforcing authority may allow a shorter period, eg in an emergency where there is a serious risk to the health and safety of any person. This shorter period is known as a 'waiver' or dispensation. You must normally notify each individual job to the enforcing authority (see Table 2 for information about which enforcing authority you should send the notification to).

119 You can use form FOD ASB5 for notification, available from the HSE website, local HSE offices or from the Asbestos Licensing Unit.

Guidance

9

120 You are required to inform the enforcing authority in writing if there are changes to the work that might affect the particulars of the notification. Table 2 tells you which enforcing authority to notify.

121 Your notification will need to include:

(a) your name and address and the address and telephone number of your usual place of business;

(b) a brief description of –

(i) the location of the worksite;

(ii) the type(s) of asbestos to be used or handled (classified in accordance with regulation (2));

(iii) the maximum quantity of asbestos of each type to be held at any one time on the premises at which the work is to take place;

(iv) the activities and processes involved;

(v) the number of workers involved;

(vi) the measures taken to limit the exposure of employees to asbestos; and

(vii) the date of the commencement of the work and its duration.

Table 2 How to identify the appropriate enforcing authority

Type of premises	*Authority to notify*
(a) Shops, offices, separate catering services including: (i) Restaurants and cafes (ii) Coin-operated launderettes (iii) Sportsgrounds (iv) Entertainment, recreational and leisure activities, gyms, health clubs and therapeutic services including solaria (v) Exhibitions (vi) Church or religious meetings (vii) Hotels and boarding houses and residential accommodation including residential homes for the elderly, other than domestic premises (viii) Camping and caravan sites (ix) Wholesale and retail storage (x) Animal boarding and care establishments and zoos (xi) Tyre and exhaust replacement/repair premises (xii) Garden centres (xiii) Child care, playgroups, nurseries (xiv) Undertakers	Unitary authority, district council (or equivalent)

Guidance

Type of premises	*Authority to notify*
(b) (i) Domestic premises (ii) Factories and factory offices (iii) Civil engineering, construction and demolition sites (iv) Hospitals (v) Research and development establishments (vi) Local government services and educational establishments (vii) Fairgrounds (viii) Radio, television and film broadcasting (ix) Sea-going ships (x) Docks (xi) Transport undertakings (xii) Farms (and associated activities) (xiii) Horticultural premises and forestries (xiv) Quarries	HM Inspector of the Health and Safety Executive
(c) Mines	HM Inspector of Mines, Health and Safety Executive
(d) Railways	HM Railway Inspector, Office of Rail Regulation
(e) Operating Licence Nuclear Sites	HM Inspector of Nuclear Installations, Health and Safety Executive
(f) Offshore Installations	Offshore Safety Division, Health and Safety Executive

122 You can notify work by telephone to the enforcing authority especially if you are not sure which authority to notify. But you must follow this up by confirmation in writing or on form FOD ASB5 to the relevant enforcing authority at least 14 days before the work starts.

9

123 You may be allowed to submit a single notification of licensable asbestos work to the enforcing authority for work which is likely to be regularly repeated on your premises. (Note: if you have several premises, you will need a separate notification for each premises.) You will also need to notify separately any other work you plan to carry out which is not covered in the original notification.

Regulation 10

Information, instruction and training

Regulation

(1) Every employer shall ensure that adequate information, instruction and training is given to those of his employees –

(a) who are or who are liable to be exposed to asbestos, or who supervise such employees, so that they are aware of –

(i) the properties of asbestos and its effects on health, including its interaction with smoking,

(ii) the types of products or materials likely to contain asbestos,

(iii) the operations which could result in asbestos exposure and the importance of preventive controls to minimise exposure,

(iv) safe work practices, control measures, and protective equipment,

10

(v) the purpose, choice, limitations, proper use and maintenance of respiratory protective equipment,

Regulation

(vi) emergency procedures,

(vii) hygiene requirements,

(viii) decontamination procedures,

(ix) waste handling procedures,

(x) medical examination requirements, and

(xi) the control limit and the need for air monitoring,

in order to safeguard themselves and other employees; and

(b) who carry out work in connection with the employer's duties under these Regulations, so that they can carry out that work effectively.

(2) The information, instruction and training required by paragraph (1) shall be –

(a) given at regular intervals;

(b) adapted to take account of significant changes in the type of work carried out or methods of work used by the employer; and

(c) provided in a manner appropriate to the nature and degree of exposure identified by the risk assessment, and so that the employees are aware of –

(i) the significant findings of the risk assessment, and

10 *(ii) the results of any air monitoring carried out with an explanation of the findings.*

ACOP

124 There are three main types of information, instruction and training (simply referred to as training from now on). These are:

(a) Asbestos awareness training. This is for those persons who are liable to disturb asbestos while carrying out their normal everyday work, or who may influence how work is carried out, such as:

(i) general maintenance staff;

(ii) electricians;

(iii) plumbers;

(iv) gas fitters;

(v) painters and decorators;

(vi) joiners;

(vii) plasterers;

(viii) demolition workers;

10 (ix) construction workers;

ACOP

(x) roofers;

(xi) heating and ventilation engineers;

(xii) telecommunications engineers;

(xiii) fire and burglar alarm installers;

(xiv) computer installers;

(xv) architects, building surveyors and other such professionals;

(xvi) shop fitters.

(b) Training for non-licensable asbestos work. This is for those who undertake planned work with asbestos which is not licensable such as a roofer or demolition worker removing a whole asbestos cement sheet in good condition or analytical staff and asbestos surveyors.

(c) Training for licensable work with asbestos. This is for those working with asbestos which is licensable such as removing asbestos insulation or insulating board.

10

Guidance 10

125 Employers should consult safety representatives and elected representatives of employee health and safety in good time about the information, instruction and training which they intend to provide.

ACOP

Asbestos awareness training

126 Asbestos awareness training is required to be given to employees whose work could foreseeably expose them to asbestos. In particular, it should be given to all demolition workers and those workers in the refurbishment, maintenance and allied trades where it is foreseeable that their work will disturb the fabric of the building because ACMs may become exposed during their work. Exemption from this requirement would apply only where the employer can demonstrate that work will only be carried out in or on buildings free of ACMs. This information should be available in the client's asbestos management plan.

127 This training should cover the following topics in appropriate detail, by means of both written and oral presentation, and by demonstration as necessary:

(a) the properties of asbestos and its effects on health, including the increased risk of lung cancer for asbestos workers who smoke;

(b) the types, uses and likely occurrence of asbestos and ACMs in buildings and plant;

(c) the general procedures to be followed to deal with an emergency, for example an uncontrolled release of asbestos dust into the workplace; and

(d) how to avoid the risks from asbestos, for example for building work, no employee should carry out work which disturbs the fabric of a building unless the employer has confirmed that ACMs are not present.

Training for non-licensable asbestos work

128 Persons requiring this type of training would include those whose work will knowingly disturb ACMs, such as maintenance workers and their supervisors; and

10

ACOP

those who carry out asbestos sampling and analysis. It should be given in addition to the asbestos awareness training outlined in paragraph 127.

129 This training should cover the following topics in appropriate detail, by means of both written and oral presentation, and by demonstration:

(a) the operations which could result in asbestos exposure and the importance of preventive controls to minimise exposure;

(b) how to make suitable and sufficient assessments of the risk of exposure to asbestos;

(c) the control limit, and the purpose of air monitoring;

(d) safe work practices, control measures, and protective equipment including an explanation of how the correct use of control measures, protective equipment and work methods can reduce the risks from asbestos, limit exposure to workers and limit the spread of asbestos fibres outside the work area;

(e) the maintenance of control measures, including where relevant the maintenance of enclosures;

(f) procedures for recording, reporting and correcting defects;

(g) the appropriate purpose, choice and correct selection from a range of suitable RPE including any limitations;

(h) the correct use, and where relevant, cleaning, maintenance and safe storage of RPE and PPE, in accordance with the manufacturer's instructions and information;

(i) the importance of achieving and maintaining a good seal between face and RPE, the relevance of fit tests, and the importance of being clean-shaven;

(j) hygiene requirements;

(k) decontamination procedures;

(l) waste handling procedures;

(m) emergency procedures;

(n) which work requires an HSE licence;

(o) an introduction to the relevant Regulations, Approved Codes of Practice and guidance that apply to asbestos work and other Regulations that deal with the carriage and disposal of asbestos;

(p) for analysts, personal sampling and leak and clearance sampling techniques; and

(q) other work hazards including working at height, electrical, slips, trips and falls.

130 Where any employees are required to use the following plant and equipment or carry out the following work activities then practical training (ie giving someone the opportunity to try and practice something for themselves rather than
10 having it explained or demonstrated to them) should be given:

ACOP

(a) use of decontamination facilities;

(b) use of PPE, particularly RPE;

(c) construction of mini-enclosures where necessary; and

(d) use of control techniques, such as Class H vacuum cleaners (BS EN 60335).

131 The procedures for providing information, instruction and training should be clearly defined and set out in a written health and safety policy document. This should be reviewed regularly, particularly when work methods change. Records should be kept of the training undertaken by each individual.

10

Guidance

Training for licensable asbestos work

132 Chapter 4 of HSG247 *Asbestos: The licensed contractors' guide (The licensed contractors' guide)*[13] sets out the detailed content of the asbestos training modules for operatives, supervisors, managers, directors, supervisory licence holders and licensed scaffolders that are involved in licensable work.

133 All information, instruction and training given should include an appropriate level of detail, suitable to the job, and should use written materials, oral presentation and demonstration as necessary.

10

ACOP

134 The following is a list of the information, instruction and training that should be given to all employees, including operatives, supervisors, managers, directors and supervisory licence holders in addition to the asbestos awareness training outlined in paragraph 127:

(a) the health risks to employees' families and others which could result from taking home contaminated equipment and clothing, its interaction with smoking and the increased risk of lung cancer for asbestos workers who smoke;

(b) the assessment of risk and the purpose of the plan of work;

(c) the operations which could result in asbestos exposure and the importance of preventive controls to minimise exposure;

(d) the control limit, the assessment of exposure and the purpose and importance of air monitoring to check compliance with the limit, including the purpose of personal sampling;

(e) safe work practices, control measures, and protective equipment including an explanation of how the correct use of control measures, protective equipment and work methods can reduce the risks from asbestos, limit exposure to workers and limit the spread of asbestos fibres outside the work area;

(f) the importance of following (and for managers and supervisors ensuring the workforce follow) the procedures, controls and preventative measures set out in the plan of work and risk assessment;

(g) the maintenance of control measures, including where relevant the maintenance of enclosures and negative pressure equipment;

(h) procedures for recording, reporting and correcting defects;

10

ACOP

(i) the appropriate purpose, choice and correct selection from a range of suitable RPE including any limitations;

(j) the correct use, cleaning, maintenance and safe storage of RPE, with specific attention to ensuring that the RPE is working correctly in accordance with the manufacturer's instructions and information;

(k) the importance of achieving and maintaining a good seal between face and RPE, the relevance of fit tests, and the importance of being clean-shaven;

(l) the suitability, correct use, storage and maintenance of protective clothing, including clothing used for transit;

(m) hygiene requirements;

(n) decontamination procedures, particularly within enclosures, airlocks (including bag locks) and hygiene units;

(o) site set-up: marking out the work area, setting up barriers, transit routes and waste storage area, pre-cleaning, sealing sources of potential leaks, construction and layout of the enclosure including negative pressure units, viewing panels and airlocks, positioning of decontamination units, air management and leak testing;

(p) controlled removal techniques and how they work including types of wet surfactant injection of sprayed asbestos and lagging, spray wetting of AIB and asbestos cement, wrap-and-cut, and (if relevant) use of glovebags;

(q) waste-handling procedures including bagging, storage and disposal;

(r) site clean-up and clearance procedures, including the certificate of reoccupation arrangements;

(s) emergency procedures including general procedures such as the uncontrolled release of asbestos fibres into the workplace or outbreak of fire;

(t) medical examination requirements;

(u) the results of any air monitoring carried out with an explanation of the findings;

(v) for analysts, personal sampling and leak and clearance sampling techniques;

(w) other work hazards including working at height, electrical, slips, trips and falls; and

(x) an introduction to the relevant Regulations, Approved Codes of Practice and guidance that apply to asbestos work and other Regulations that deal with the carriage and disposal of asbestos.

135 To assist the employer to comply with their legal duties under the Control of Asbestos Regulations, the following additional training should be given to supervisors, managers, directors and supervisory licence holders, at an appropriate level, so that they can effectively carry out their role on site. This should include:

10 (a) their responsibilities for directing, supervising and monitoring all aspects of work on site, including people's health and safety;

ACOP

(b) the importance of the supervisor being on site at all key stages of the work (witnessing the smoke test, ensuring that the hygiene facilities are fully operational before work starts, ensuring signs and barriers are correctly erected, carrying out daily checks) to ensure that it is done safely;

(c) how to produce and apply plans of work that set out the appropriate procedures, controls and preventative measures based on the assessment, including how and when to update plans;

(d) how and when to notify the appropriate enforcing authorities that work is taking place and situations where re-notification is necessary;

(e) how to deal with situations where the methods set out in the plan of work cannot be followed due to a change in circumstances and a revision to the plan is required;

(f) the application of suitable contingency procedures in the event of a failure of controls;

(g) the importance of monitoring and auditing the work activities;

(h) the importance of having effective arrangements in place to communicate with and monitor workers inside the enclosure and hygiene unit;

(i) a need to provide additional training, information and instruction to workers as necessary such as the use of a particular piece of equipment or work method for which training has not previously been given;

(j) how to assess the competence of employees and identify their training needs;

(k) when and how air monitoring should be undertaken, how the results are interpreted and to whom they should be communicated;

(l) how the results and records of personal air sampling, fit tests and medicals should be kept and maintained and to whom they should be communicated;

(m) how to apply the procedures for dealing with accidents, incidents and emergencies;

(n) keeping the work area clean and free of asbestos;

(o) the importance of ensuring that the correct procedures are followed at the end of the job to allow a certificate of reoccupation to be issued; and

(p) an understanding of what the laboratory analyst will require before clearance sampling is undertaken and the certificate of reoccupation can be issued.

136 Practical training is essential for those entering enclosures such as operatives, supervisors and supervisory licence holders. Practical training is also required where people are required to use the following plant and equipment or carry out the following work activities or procedures:

(a) decontamination procedures and use of hygiene facilities;

(b) use of PPE, particularly RPE;

10 (c) construction of enclosures, airlocks and achieving sufficient numbers of air changes within the enclosure;

ACOP

(d) controlled removal techniques, including the use of multiple and single needle injection systems, glovebags and wrap-and-cut; and

(e) waste removal procedures on site including double bagging and removal through the bag lock.

137 Anyone who carries out any examination, testing (including clearance inspection, air monitoring and exposure monitoring) or maintenance of plant or equipment (eg LEV systems and RPE) should have had sufficient training and experience in inspection methods and techniques to ensure that they are competent.

Safety representatives

138 Training for safety representatives and elected representatives of employee safety will need to be appropriate to their role.

Competence of those providing training

10 139 All training should be provided by someone who is competent to do so, who has had adequate personal practical experience and who has a theoretical knowledge of all relevant aspects of the work being carried out by the employer.

Guidance

Provision of information

140 All training certificates issued by such people or organisations should be traceable and have a validity of no more than one year. The employer should carry out checks as may be necessary to establish the authenticity of training certificates. More information on training for licensable work can be found in *The licensed contractors' guide*.[13]

10 141 For licensable work, copies of the respective training records should be provided to each individual. The original of the records should be kept centrally and be reviewed annually to help inform what refresher training is required or earlier if concerns are raised about an individual's competence.

ACOP

142 Employers should make the following information available to employees and safety representatives:

(a) a copy of the current assessment for the work;

(b) a copy of the plan of work;

(c) details of any air monitoring strategy and results;

(d) maintenance records for control measures;

(e) personal information from health records (only relating to the individual employee concerned);

(f) a copy of the individual's training record (only relating to the individual employee concerned);

10 (g) the results of any face-fit test for asbestos RPE.

ACOP

143 For licensable work, this information should also include:

(a) a copy of the licence;

(b) details of notification under regulation 9 made to the enforcing authority;

10 **(c) any anonymous collective information from the health records.**

Guidance

144 Where the results of air monitoring show that the relevant control limit has been unexpectedly exceeded, employers should tell employees, safety representatives and elected representatives of employee health and safety about this as quickly as possible and give details of the reasons for what happened and

10 the action taken or proposed.

ACOP

Duration of training

145 The duration of training should be appropriate to the type of training (initial training or refresher training), the role for which the person is being trained (operative, supervisor or manager) and the nature of the work being trained for (non-licensable work, asbestos removal, work ancillary to asbestos removal eg

10 **scaffold work, maintenance of plant and equipment etc).**

Guidance

10 146 Further guidance on course durations for licensable work can be found in *The licensed contractors' guide*.

ACOP

Refresher training

147 Refresher training should be given at least every year and should be appropriate to the role undertaken. Those persons who require only awareness training could have refresher training as part of other health and safety updates. Employers should identify the specific training needs of their employees so that the refresher training can be appropriately tailored. It should not be a repeat of the initial training. Where training needs dictate, refresher training should include an appropriate element of practical training, particularly covering decontamination procedures, use of RPE, and controlled removal techniques. Refresher training will be required more frequently than annually if:

(a) work methods change;

(b) the type of equipment used to control exposure changes; or

10 **(c) the type of work carried out changes significantly.**

Guidance

148 Refresher training should be appropriate to the role, eg licensable work or non-licensable work. Supervisors of licensable work, for example, should receive supervisory refresher training, not operative refresher training and it should be appropriately tailored. Refresher training should include reviewing where things

10 have gone wrong and sharing good practice.

ACOP

Training of non-employees

149 Employers have a duty under regulation 3(4)(a) of the Asbestos Regulations to ensure, so far as is reasonably practicable, that adequate information, instruction and training is given to non-employees who are on the premises and

10 **could be affected by the work, as well as to their own employees.**

ACOP 10

150 This will need to take account of any possible risks resulting from rearrangement of thoroughfares and fire exits as well as of the risks arising from disturbance of ACMs.

Guidance 10

151 For those entering enclosures in emergency situations, please refer to paragraph 251.

Procedures for providing information, instruction and training

152 The procedures for providing information, instruction and training for licensable work should be clearly defined and set out in a written document. This should be reviewed regularly, particularly when work methods change. Records should be kept of the training undertaken by each individual.

153 For licensable work, copies of the respective training records should be provided to each individual. The originals of the records should be kept centrally, and be reviewed annually to help inform what refresher training is required, or earlier if concerns are raised about an individual's competence after an inspection or incident.

ACOP 10

154 Employers should consult safety representatives and elected representatives of employee safety in good time about the information, instruction and training which they intend to provide.

Regulation 11 Prevention or reduction of exposure to asbestos

Regulation 11

(1) Every employer shall –

(a) prevent the exposure of his employees to asbestos so far as is reasonably practicable;

(b) where it is not reasonably practicable to prevent such exposure –

(i) take the measures necessary to reduce the exposure of his employees to asbestos to the lowest level reasonably practicable by measures other than the use of respiratory protective equipment, and

(ii) ensure that the number of his employees who are exposed to asbestos at any one time is as low as is reasonably practicable.

(2) Where it is not reasonably practicable for the employer to prevent the exposure of his employees to asbestos in accordance with paragraph (1)(a), the measures referred to in paragraph (1)(b)(i) shall include, in order of priority –

(a) the design and use of appropriate work processes, systems and engineering controls and the provision and use of suitable work equipment and materials in order to avoid or minimise the release of asbestos; and

(b) the control of exposure at source, including adequate ventilation systems and appropriate organisational measures,

and the employer shall so far as is reasonably practicable provide the employees concerned with suitable respiratory protective equipment in addition to the measures required by sub-paragraphs (a) and (b).

Regulation

(3) Where it is not reasonably practicable to reduce the exposure of an employee to asbestos to below the control limit by the measures referred to in paragraph (1)(b)(i), then, in addition to taking those measures, the employer shall provide that employee with suitable respiratory protective equipment which will reduce the concentration of asbestos in the air inhaled by the employee (after taking account of the effect of that respiratory protective equipment) to a concentration which is –

(a) below the control limit; and

(b) is as low as is reasonably practicable.

(4) Personal protective equipment provided by an employer in accordance with this regulation or with regulation 14(1) shall be suitable for its purpose and shall –

(a) comply with any provision of the Personal Protective Equipment Regulations 2002[(a)] which is applicable to that item of personal protective equipment; or

(b) in the case of respiratory protective equipment, where no provision referred to in sub-paragraph (a) applies, be of a type approved or shall conform to a standard approved, in either case, by the Executive.

(5) The employer shall –

(a) ensure that no employee is exposed to asbestos in a concentration in the air inhaled by that worker which exceeds the control limit; or

(b) if the control limit is exceeded –

(i) forthwith inform any employees concerned and their representatives and ensure that work does not continue in the affected area until adequate measures have been taken to reduce employees' exposure to asbestos to below the control limit,

(ii) as soon as is reasonably practicable identify the reasons for the control limit being exceeded and take the appropriate measures to prevent it being exceeded again, and

(iii) check the effectiveness of the measures taken pursuant to sub-paragraph (ii) by carrying out immediate air monitoring.

11

(a) S.I. 2002/1144

ACOP

11

155 Work which disturbs ACMs should only be carried out when there is no other reasonably practicable way of doing the work or the alternative method creates a more significant risk. Employers must therefore first decide whether they can prevent the exposure to asbestos so far as is reasonably practicable, before considering how they will reduce the exposure to as low as reasonably practicable.

Guidance

11

156 It may be that the work which disturbs the asbestos or ACM is not necessary or that a method can be chosen which does not involve disturbing the asbestos, or minimises such disturbance, such as:

Guidance 11

(a) re-routing cables away from ACM; or

(b) covering up materials rather than working on them.

ACOP

157 Where it is not reasonably practicable to prevent exposure, it must first be reduced to the lowest level reasonably practicable by means other than the use of RPE.

158 When it is not reasonably practicable to prevent exposure to asbestos the employer must choose the most effective method or combination of methods which minimises fibre release and thereby reduces the exposure to the lowest levels reasonably practicable and document this in the written risk assessment/plan of work. Such work methods could include the following:

(a) removing materials containing asbestos before any other major work (such as demolition or refurbishment) begins using the most effective methods that minimise fibre release, eg controlled wet stripping techniques;

(b) choosing methods that do not involve dry working and avoiding abrasion, sanding, machining or cutting etc of ACMs;

(c) choosing work methods which present the least overall risk;

(d) where necessary carrying out a pre-clean of the work area before removal work takes place;

(e) keeping the work area clean by clearing up at regular intervals, using methods which do not spread asbestos dust, so that waste debris and dust do not accumulate and by not allowing waste to be kept on site for longer than necessary. Vacuum-cleaning equipment of Class H (BS EN 60335[9]) should be used to clean up any dust and debris which may contain asbestos;

(f) removal of a whole door instead of an AIB panel attached to it;

(g) wrapping and cutting off or removing pipes at the flange joints, rather than disturbing the insulation material on them;

(h) re-routing cables and services away from ACM;

(i) protecting ACMs from damage when working near them; and

11 **(j) cutting the bolts of asbestos cement sheets from a safe working platform and removing them whole.**

Guidance 11

159 Note that asbestos cement is highly fragile and you should **never** walk on it. If you need to work on or over asbestos cement, a safe system of work must be devised and used.

ACOP

160 Employers should keep the number of employees exposed, and others who might be exposed, to asbestos at any one time as low as reasonably practicable.

11 **161 Airborne levels should be reduced to as low a level as reasonably practicable and exposure should be controlled so that any peak exposure is less than 0.6 fibres per cm^3 averaged over a maximum continuous period of 10 minutes by the use of appropriate RPE if exposure can not be reduced sufficiently by other means.**

Guidance

11

162 It will usually be reasonably practicable to achieve much lower exposures by the proper use of RPE. Work which disturbs ACMs should only be carried out when there is no other reasonably practicable way of doing the necessary work, or the alternative method creates a more significant risk, or to reduce the risk of exposure to asbestos (eg from friable ACMs in poor condition).

ACOP

11

163 A sufficient number of suitable viewing panels should be installed in enclosures. The panels will allow managers to monitor the work of their employees without being unnecessarily exposed. The viewing panels should be located to ensure that all areas inside the enclosure are visible as far as possible.

Guidance

11

Non-licensable asbestos removal work

164 The duty to avoid, or if that is not reasonably practicable, adequately control exposure, applies equally to work with asbestos for which a licence is not needed.

165 Companies should have in place a policy always to check whether ACMs are present before carrying out work which disturbs the fabric of a building which may contain asbestos. The policy should also ensure that work which does disturb ACMs is restricted to authorised people who have been given the necessary information, instruction and training. In addition, non-licensed asbestos work should be carried out using the most effective method or combination of methods which minimises fibre release and thereby reduces the exposure to the lowest levels reasonably practicable. Some examples of work methods which reduce dust emission are:

(a) removing sheets whole;

(b) avoiding the inappropriate use of power tools;

(c) keeping materials thoroughly wet; and

(d) utilising LEV systems such as cowls on drills and by shadow vacuuming.

166 For further guidance on methods of working with asbestos cement,[14,15] minor works with building materials containing asbestos and on work with asbestos-containing textured decorative coatings, refer to the *Asbestos essentials task manual.*[10]

ACOP

11

Work on textured decorative coatings containing asbestos

167 Textured decorative coatings can be worked on or disturbed for a number of reasons, including during maintenance and repair, during replacement of lights and other fittings, during removal as part of room or building refurbishment and during removal or repair following damage due to water, fire or other accidental or deliberate acts.

168 As indicated in paragraphs 36-37, most work with textured decorative coatings is likely only to produce sporadic and low intensity worker exposure and can be categorised as complying with regulation 3(2) as long as 3(2)(b) is fulfilled, ie it is clear from the risk assessment that the control limit will not be exceeded.

169 When deciding what controls are needed for any particular job, due regard should be given to the:

(a) risk assessment for the work (see paragraphs 53-69);

ACOP

(b) nature and extent of the material and of the planned work;

(c) proposed method of removal;

(d) potential risks from materials other than asbestos.

170 The following apply to all work with textured decorative coatings whether the exceptions detailed in paragraph 31 apply or not:

(a) the duty to prevent exposure to asbestos as far as is reasonably practicable, or, if that is not possible to control it as far as is reasonably practicable (see paragraphs 155-165 and 167-169);

(b) in line with (a) above, the method of work should aim to minimise release of asbestos fibres;

(c) sanding, particularly with power tools, must be avoided;

(d) work must only be carried out by suitably trained and competent workers (see paragraphs 128-131);

(e) a suitable plan of work must be prepared before the work starts (see paragraphs 76-82); and

(f) suitable RPE must always be worn (see paragraph 175(a)).

171 For very minor work, it may not - subject to a suitable and sufficient risk assessment - be reasonably practicable to use the degree of control required for more extensive work. For example, shadow vacuuming with a HEPA vacuum and suitable RPE is likely to be sufficient for work which requires the removal or insertion of a few screws through textured coating.

172 More significant work usually produces substantial amounts of other waste such as paint flakes and plaster debris and will require more stringent controls than minor work.

173 The control regime for all but the most minor work should include control at source/dust suppression and good work practices which minimise dust generation and spread (avoiding power tools etc). The primary controls will consist of one or more of the following:

(a) remove whole underlying plasterboard if possible with textured decorative coatings attached;

(b) steam to loosen (there are proprietary machines/equipment available) and scrape;

(c) apply a hydrating gel to loosen and scrape;

(d) apply 'solvent free' chemical and scrape.

174 In certain circumstances appropriate wet blasting techniques may also be necessary for residual sections or very stubborn material. These techniques should not be employed as the primary method of removal.

175 The following measures should be employed to help contain and prevent the spread of asbestos fibres and other materials during all but the most minor work with textured coatings:

11

ACOP

(a) Although worker exposure is unlikely to exceed the control limit during controlled removal, RPE should always be worn. In line with expected asbestos airborne dust levels suitable RPE such as disposable (FFP3) or half-mask respirators (with P3 filters) should be adequate. Other suitable PPE, eg coveralls should also be worn.

(b) In situations where the work on textured decorative coatings is necessary due to water, fire or other accidental or deliberate damage, the area is likely to contain debris and other waste materials (although the occupier may already have taken steps to clean the area(s)). The affected area(s) should be inspected to identify the extent of debris. If there is any evidence of textured decorative coating debris or dust then a pre-clean should be undertaken. The enclosure and protective sheeting should be installed as far as possible before the pre-clean is started. The pre-clean should be undertaken using appropriate dust suppression and control measures including vacuuming with a Class H vacuum cleaner (BS EN 60335[9]), surface wiping, temporary encapsulation with tape, spray wetting and bagging (see also paragraphs 247 and 249). Once all the contamination has been removed, the enclosure and protective sheeting installation should be completed and the main work can then start.

(c) The work area should be segregated and enclosed using the existing room or a purpose-built enclosure. Access to the enclosure should be regulated through a 2-stage airlock. Workers should decontaminate in the airlock system prior to leaving the work area. This will involve vacuuming down using a Class H vacuum cleaner (BS EN 60335[9]) and washing footwear and wiping the RPE in the inner stage. PPE and RPE should be removed in the outer stage of the airlock.

(d) Any portable items liable to become contaminated with dust and debris from textured decorative coatings should be removed prior to work starting. Remaining items should be protected with plastic sheeting.

(e) Once the work is complete, the area should be thoroughly cleaned before being returned to the occupier. All visible traces of dust and debris should be removed and the enclosure dismantled. A thorough visual inspection should then be carried out. Clearance air monitoring is not necessary as part of the clearance procedures.

(f) An independently provided certificate of reoccupation is not necessary but a written statement should be issued to the occupier stating that the area has been thoroughly cleaned and visually inspected to ensure that no visible traces of dust and debris remain and is suitable for reoccupation. The statement should also contain the following information: the site address, the dates of the work, a brief description of the work, the name and address of the contractor, and details of the specific areas and items visually checked. The statement should clearly bear the name and signature of the person completing the inspection.

176 There are certain practical difficulties in removing textured decorative coatings from damaged ceilings. The 'usual' practice is to 'pull-down' or 'drop' the ceiling which can be a dusty process (the dust is predominantly calcium sulphate or calcium silicate). In these situations there should be greater attention paid to reducing general dust levels (a requirement under the COSHH Regulations[16]). The ceiling (boards or lathe and plaster) should be thoroughly dampened down using an effective wetting system before work starts and during the work. Material, debris and dust on the floor should also be dampened down.

11

Guidance 11

177 If work methods in paragraphs 171-176 are complied with it can be deemed that the control limit will not be exceeded and regulation 3(2) will apply. It will only be necessary to provide additional substantiation under the risk assessment (in accordance with paragraphs 64 and 67(a)) if there is significant or exceptional variance in the scope or method of work.

ACOP 11

Licensable asbestos removal work

178 Employers must choose the methods which are most effective at reducing fibre release at source.

Removal of insulation and coating

179 For work with asbestos insulation and coating, this usually means controlled wet stripping and avoiding the use of abrasive power tools.

Guidance 11

180 The standards in British Standards Institution (BSI) PAS 60, or better, should be followed. Wet injection equipment purchased after 2005 should conform to PAS 60-1:2004.[17]

181 There may, however, be situations where other techniques such as wrap and cut may be more efficient at preventing or reducing exposure (eg for the removal of redundant pipework).

ACOP 11

182 Wet injection techniques should uniformly wet the asbestos material before its removal. The wetting agent will need sufficient time to thoroughly penetrate the ACM but oversaturation should be avoided. Oversaturation can lead to pieces falling off and the formation of pools of liquid which may turn the material into unmanageable slurry. It is essential that employers check the degree of saturation, eg by visual inspection, examining texture or by using dyes, before attempting removal. The treated insulation should be of a dough-like consistency.

183 Where the ACM is being removed from its substrate, employers must not use dry stripping methods unless there is no reasonably practicable alternative (stripping using a glovebag without any form of wetting is a dry method). Where it is considered that there is absolutely no alternative to dry stripping methods, this must be justified by the risk assessment and clearly detailed in the plan of work. Employers must make sure that effective measures are used to control fibre release in the work area (eg glovebags within the enclosure, vacuum transfer).

Guidance 11

184 Before agreeing to a client's request for the work to be carried out dry, contractors should discuss the request with the enforcing authority.

ACOP 11

185 Occupiers and owners of buildings and plant should co-operate with contractors undertaking the work, releasing plant wherever practicable so that it can be isolated and worked on cold and free from electrical and chemical risks.

Work with asbestos insulating board (AIB)

186 As indicated in paragraph 41, certain work with AIB that is a short, non-continuous maintenance activity and which conforms to the principles of good practice can be considered to conform to regulation 3(2) and will be non-licensable. For any work with AIB, employers must use work methods which reduce the level of fibres released at source, wherever possible by removing boards whole without breaking them and, wherever practicable, by controlled stripping techniques including spraying with wetting agents and using Class 'H' vacuum cleaners (BS EN 60335[9]).

Guidance 11

187 More guidance on fibre control methods can be found in *The licensed contractors' guide*[13] and, for non-licensable work with AIB, in the *Asbestos essentials task manual*.[10]

ACOP

Work maintaining plant and equipment contaminated with asbestos

188 All maintenance of equipment contaminated with asbestos and where there is a risk of fibre release must be done under controlled conditions. For example when the item of plant needs to be stripped down (eg air extraction equipment), it should be carried out within a permanently set aside work area which is under negative pressure and which is connected to the hygiene facilities by an airlock system. Where this work is done on site then it must be done within an enclosure.

189 In addition, dust should be controlled at source by using local exhaust ventilation (LEV) or controlled wetting techniques, eg airless or low-pressure spraying. Care should be taken not to overwater when working on or near electrical equipment. Alternatively, dielectric fluids should be used. Those who carry out such maintenance work either on site or on the maintenance company's own premises will require a licence for ancillary work.

Respiratory protective equipment

190 If from the assessment of the work it is concluded that, despite the use of other control measures, the exposure of an employee is liable to exceed the control limit or exceed 0.6 f/cm^3 peak level measured over 10 minutes, the employer must provide suitable RPE (see regulation 11(3)) to reduce exposure to a level as low as is reasonably practicable and which must be below the control limit. In addition, the employer should make sure that the RPE is used correctly by those carrying out the work.

11

191 The control limit also triggers the need for immediate steps to be taken under regulation 11(5) and for respirator zones which are required under regulation 18.

Guidance 11

192 RPE should be worn where there is any work which could disturb asbestos. Employers must choose RPE which is designed to protect against exposures well above those expected, to allow for unexpected high exposures and to provide an adequate margin of safety. If there is doubt about the level of protection, employers must always select higher performance equipment, provided that it is suitable for the work being carried out. Employers must make sure that RPE is used correctly by those carrying out the work.

ACOP 11

193 To be suitable, RPE must be matched to the job, the environment, the anticipated maximum exposure, and the wearer, and take into account such issues as facial hair and spectacles. It should be compatible with any other personal protective equipment (PPE). In particular, any PPE which protects the head or eyes of employees should not affect the fit of the RPE.

Guidance 11

194 For licensable work inside enclosures, power-assisted full facepiece respirators fitted with P3 filters should be worn. Alternatively similar or higher performance equipment can be used, eg powered hoods or blouse or air-fed equipment. Disposable RPE (eg FFP3) or half-mask RPE (with P3 filters) can be used in low-risk ancillary tasks as identified in the risk assessment. These situations could include scaffold erection, site set-up, enclosure dismantling, and waste handling outside the enclosure.

Guidance

11

Fit-testing of face pieces

195 The performance of RPE with a tight-fitting face piece (ie filtering face pieces, half and full face masks) depends on a good contact between the wearer's skin and the face seal of the mask. A good face seal can only be achieved if the wearer is clean-shaven in the region of the seal and the face piece is of the correct size and shape to fit the wearer's face. If spectacles with side arms are worn together with PPE, they should not interfere with the correct fitting of the face piece or the face seal. The performance of RPE with a loose-fitting face piece (eg visors, helmets, hoods) is less dependent on a tight fit on the face, but nevertheless requires the correct size to ensure the wearer achieves an adequate fit and protection.

ACOP

11

196 Employers should ensure that the selected face piece (tight and loose fitting types) is of the right size and can correctly fit each wearer. For a tight-fitting face piece (ie filtering face pieces usually known as disposable masks, half and full face masks) the initial selection should include fit-testing to ensure the wearer has the correct device. The test will assess the fit by determining the degree of face seal leakage of a test agent while the RPE user is wearing the face piece under test. For full face masks, a suitable quantitative fit-test should be used and the pass level fit factor is 2000. For devices such as filtering face pieces and half masks, the pass level fit factor is 100. For these lower performance face pieces, a suitable and validated qualitative method (often called a semi quantitative test) or the quantitative fit test can be used. Employers must ensure that whoever carries out the fit-testing is competent to do so.

197 Repeat fit-testing will be needed when changing to a different model of RPE or a different sized face piece or if there have been significant changes to the facial characteristics of the individual wearer, eg as a result of significant weight gain or weight loss or due to dentistry.

Guidance

11

198 It is good practice to have a system in place to ensure repeat fit-testing of RPE is carried out on a regular basis, eg annual testing for workers involved in licensed asbestos removal.

199 If an employee changes jobs, employers should carry out such checks as may be necessary to establish the authenticity of fit-testing certificates provided by employees.

200 The quantitative fit-testing may be carried out using:

(a) a test chamber which uses a salt aerosol or sulphur hexafluoride gas to assess the face seal leakage;

(b) a portable device at the workplace which measures particulates in air to assess the face seal leakage; or

(c) a portable device at the workplace which measures pressure variations inside the face piece to assess face seal leakage.

201 Qualitative test methods use bitter or sweet-tasting aerosols. When the tests are carried out the facepiece wearer will perform simple exercises as part of the test protocol. More information on the selection, including information on assigned protection factors, use and fit-testing of RPE is contained in *Fit testing of respiratory protective equipment facepieces*,[18] and in *The licensed contractors' guide*.[13]

Regulation 12

Use of control measures etc

Regulation

(1) Every employer who provides any control measure, other thing or facility pursuant to these Regulations shall take all reasonable steps to ensure that it is properly used or applied as the case may be.

(2) Every employee shall make full and proper use of any control measure, other thing or facility provided pursuant to these Regulations and, where relevant, shall –

(a) take all reasonable steps to ensure that it is returned after use to any accommodation provided for it; and

(b) if he discovers a defect therein report it forthwith to his employer.

12

ACOP

Control measures

202 Employers should have procedures in place to make sure that control measures are properly used or applied and are not made less effective by other work practices or other machinery. These procedures should include:

(a) regular checks, at least at the start of every shift; and

(b) prompt action when a problem is identified.

General duties on employees

203 Within the general duties imposed by regulation 12(2), employees should, in particular:

(a) use any control measures, including dust suppression and extraction equipment and RPE, and protective clothing properly and keep it in the places provided;

(b) follow carefully all the procedures set out in the employer's assessment and plan of work, including those for changing and decontamination, and comply with the use of control measures;

(c) keep the workplace clean;

(d) eat, drink and smoke only in the places provided; and

(e) report any defects concerning control measures to their supervisor/manager immediately.

204 RPE should never be taken off and put down in a contaminated area, except in the case of a medical emergency. When not in use, RPE should not be hung around the neck or in any other way be allowed to come into contact with contaminated clothing. It should not be stored in a contaminated area. RPE and protective clothing should be removed at the end of the working period, cleaned (see paragraph 221 for RPE and paragraphs 236-243 for PPE) and then placed in storage provided specifically for that purpose which is clean and will protect it from damage. Disposable RPE and protective clothing once used should be treated as asbestos waste. Before it is used, disposable RPE should be kept in a suitable container to keep it free from contamination.

12

Regulation 13

Maintenance of control measures etc

Regulation

(1) Every employer who provides any control measure to meet the requirements of these Regulations shall ensure that –

(a) in the case of plant and equipment, including engineering controls and personal protective equipment, it is maintained in an efficient state, in efficient working order, in good repair and in a clean condition; and

(b) in the case of provision of systems of work and supervision and of any other measure, it is reviewed at suitable intervals and revised if necessary.

(2) Where exhaust ventilation equipment or respiratory protective equipment (except disposable respiratory protective equipment) is provided to meet the requirements of these Regulations, the employer shall ensure that thorough examinations and tests of that equipment are carried out at suitable intervals by a competent person.

(3) Every employer shall keep a suitable record of the examinations and tests carried out in accordance with paragraph (2) and of repairs carried out as a result of those examinations and tests, and that record or a suitable summary thereof shall be kept available for at least 5 years from the date on which it was made.

13

ACOP

Maintenance of control measures

205 When working with asbestos, employers should make sure that maintenance procedures are drawn up for all control measures and for PPE. These should include the equipment used for cleaning, the washing and changing facilities, and the controls to prevent the spread of contamination. The procedures should make clear:

(a) which control measures require maintenance;

(b) when and how the maintenance is to be carried out; and

(c) who is responsible for maintenance and for making good any defects.

13

Guidance

Maintenance of enclosures

206 For licensable work, enclosures are normally required to prevent the spread of asbestos and prevent the exposure of people other than employees who may be affected by the work. Enclosures may also be required for non-licensed work but their requirements may vary slightly, eg there is no requirement for extraction in an enclosure used for textured decorative coating removal.

13

ACOP

207 If an enclosure is being used, then when work has started employers should make sure that:

(a) the enclosure is properly maintained;

(b) a thorough visual inspection of the integrity of the enclosure, airlocks and the ducting from the air extraction equipment is carried out at least at the beginning of each shift;

13

ACOP

(c) air extraction equipment is operated in the enclosure throughout the work, including breaks, and for at least 60 minutes after the end of each shift;

(d) air monitoring is undertaken outside the enclosure when appropriate, eg when the air exhausted from the enclosure is discharged into an occupied building because it is not reasonably practicable to discharge externally;

(e) any defect found during inspection and testing is repaired immediately;

(f) a record of inspections, tests, and defects repaired is kept available on site for inspection by the enforcing authority;

(g) viewing panels are maintained in a clean state to ensure clear visibility; and

(h) barriers are maintained on open sites.

Maintenance of hygiene facilities

208 If specific hygiene facilities are necessary such as for licensable work, then daily checks should be made throughout the duration of the work to ensure the showers, heating, lighting, extractor unit, battery-charging facilities and residual current devices are all working. The shower should provide sufficient quantities of water at a reasonable temperature and pressure to allow thorough decontamination. Blocked shower rosettes and systems which result in alternate hot and cold water are not acceptable. A record of inspections and defects repaired should be kept available on site for inspection by the enforcing authority (see Table 2).

Maintenance of vacuum cleaners

13 209 Employers should make sure that vacuum-cleaning equipment is inspected weekly when in use and is tested and examined thoroughly every six months. This requirement applies to all vacuum equipment including for non-licensed work. Before the vacuum cleaner is used, employers should ensure that the equipment is working effectively and providing adequate suction. During licensed work, the waste bag should be inspected to see whether it needs to be emptied. Due to the potential for fibres to be released, waste bags should only be inspected and changed under controlled conditions to prevent spread of contamination, ie within the enclosure, by workers wearing PPE, including RPE.

Guidance

13 210 A record of inspection, examination, maintenance and of defects remedied should be kept available for inspection by the enforcing authority. Further information on training, operation, cleaning, maintenance and record keeping for Class 'H' vacuum cleaners (BS EN 60335) can be found in PAS 60-3:2004.[19]

ACOP

Maintenance of air extraction equipment

13 211 All air extraction equipment which is necessary (which includes air movers or negative pressure units (NPUs)), including extraction units provided on hygiene facilities and relevant ventilation equipment in laboratories handling asbestos, should be visually inspected daily when in use, and should be thoroughly examined and tested at least once every six months by a competent person to make sure that it is working properly to its design specification. A record of inspection, examination, maintenance and of defects remedied must be kept available for inspection by the enforcing authority.

Guidance 13

212 Further information on design, operation, airflow testing and record keeping for negative pressure units can be found in PAS 60-2:2004.[20]

ACOP

Maintenance of wet injection equipment

13 213 All wet injection equipment should be visually inspected daily when in use, and should be thoroughly examined and tested at least once every twelve months by a competent person to make sure that it is working properly to its design specification. A record of inspection, examination, maintenance and of defects remedied must be kept available for inspection by the enforcing authority (see Table 2).

Guidance 13

214 Further information on design, operation and flow measurements for needles in wet injection equipment can be found in PAS 60-1:2004.

ACOP

Maintenance and storage of disposable respiratory protective equipment

215 Disposable RPE should always be examined in accordance with the manufacturer's instructions before it is used to ensure it is not damaged in any way and is in good working order. The pre-use examination should include checks on:

(a) the condition of the straps and face piece including the seal and the nose-piece; and

(b) the condition of the exhalation valve, if fitted.

216 The pre-use examination by the wearer should also include a fit check to ensure that the mask properly fits the wearer. The manufacturer's instructions will give information on simple fit checks, such as those involving blocking filters and inhaling to create suction inside the mask so that any leakage can be detected.

217 Disposable RPE should be stored in a suitable safe and clean location before use. A copy of the manufacturer's user instructions should be available to the wearer. Disposable RPE should be disposed of as asbestos waste after use.

Maintenance and storage of non-disposable respiratory protective equipment

218 Non-disposable RPE should be examined by a competent person to make sure that it is, and continues to be, in good working order. The RPE should be examined by a competent person before it is issued to any wearer for the first time. It should also be checked by the wearer before and after it is used to make sure that it is free from contamination and has not been damaged. The RPE should also be given a thorough examination and test by a competent person at periodic intervals (see paragraphs 222-225).

219 The pre-issue and periodic thorough examination and test should be carried out in accordance with the manufacturer's instructions. This may involve the equipment being disassembled and then inspected and tested as appropriate, before being reassembled. Each component should be visually examined in detail to ensure it is in good condition, is not damaged, cracked, broken or perished and is working properly. The examination should include the following checks:

13 (a) the condition of the head harness and face piece including the face-seal and the visor, and breathing hose if fitted;

ACOP

(b) the condition of inhalation and exhalation valves, if fitted;

(c) the condition of any threaded connections and associated gaskets/seals to ensure that they can be securely and correctly fastened;

(d) where filters are fitted, checking that they are the right type, fitted correctly, not damaged and they are within the end of shelf life printed on the filters;

(e) the battery unit including its charge and condition; and

(f) the airflow rate.

220 The pre-use test by the wearer should cover the same checks but may not involve the equipment being completely disassembled. It should also involve an inspection for any visual contamination and a fit check to ensure it is properly fitted by the wearer. The manufacturer's instructions will give information on simple fit checks, such as those involving blocking filters and inhaling to create suction inside the mask so that any leakage can be detected.

221 RPE needs to be decontaminated, cleaned and dried after each use. It should also be disinfected whenever the equipment is transferred from one person to another. RPE should be stored in a suitable safe and clean location before use. A copy of the manufacturer's user instructions should be available to the wearer.

222 Thorough maintenance, examinations and, where appropriate, tests of items of RPE should be made at least once every month, and more frequently where the health risks and conditions of exposure are particularly severe.

223 However, in situations where respirators are used only occasionally, an examination and test should be made prior to next use, and maintenance carried out as appropriate. The person who is responsible for managing the maintenance of RPE should determine suitable intervals between examinations, but in any event, the intervals should not exceed three months for equipment in use. Emergency escape type RPE should be examined and tested in accordance with the manufacturer's instructions.

13

224 A record of fit-testing, inspection, examination, maintenance and defects remedied must be kept available for five years, for inspection by the enforcing authority (see Table 2) with copies of the most recent records kept available on site.

Guidance 13

225 More guidance on RPE can be found in *The licensed contractors' guide.*[13]

Regulation 14

Provision and cleaning of protective clothing

Regulation

(1) Every employer shall provide adequate and suitable protective clothing for such of his employees as are exposed or are liable to be exposed to asbestos, unless no significant quantity of asbestos is liable to be deposited on the clothes of the employee while he is at work.

(2) The employer shall ensure that protective clothing provided in pursuance of paragraph (1) is either disposed of as asbestos waste or adequately cleaned at suitable intervals.

14

(3) The cleaning required by paragraph (2) shall be carried out either on the premises where the exposure to asbestos has occurred, where those premises are suitably equipped for such cleaning, or in a suitably equipped laundry.

Regulation

(4) The employer shall ensure that protective clothing which has been used and is to be removed from the premises referred to in paragraph (3) (whether for cleaning, further use or disposal) is packed, before being removed, in a suitable receptacle which shall be labelled in accordance with the provisions of Schedule 2 as if it were a product containing asbestos or, in the case of protective clothing intended for disposal as waste, in accordance with regulation 24(3).

(5) Where, as a result of the failure or improper use of the protective clothing provided in pursuance of paragraph (1), a significant quantity of asbestos is deposited on the personal clothing of an employee, then for the purposes of paragraphs (2), (3) and (4) that personal clothing shall be treated as if it were protective clothing.

14

ACOP

Protective clothing

226 As part of the assessment, the employer must decide whether or not protective clothing is required for work with asbestos. The assessment should start from the assumption that protective clothing will be necessary unless there is no potential for physical contamination and/or airborne exposures will be extremely slight and infrequent. For work which requires a licence, exposure is liable to be significant and employers will always need to provide a full set of PPE.

227 The protective clothing must be adequate and suitable and include footwear, whenever employees are liable to be exposed to a significant amount of asbestos debris or fibres. It should be appropriate and suitable for the job and must protect the parts of the body likely to be affected. If the assessment has concluded that a risk of contamination exists, disposable overalls (of a suitable standard fitted with a hood) and boots without laces will be required.

14

Guidance 14

228 Further PPE may be required based on the outcome of the assessment, eg waterproof clothing for outdoor work, gloves for direct hand contact with ACMs.

ACOP

Suitability of protective clothing

229 To be adequate and suitable and depending on the circumstances, the protective clothing must:

(a) fit the wearer;

(b) be of sufficient size to avoid straining and ripping the joints;

(c) be comfortable and, where appropriate, allow for the effects of physical strain;

(d) be suitable for cold environments;

(e) prevent penetration by asbestos fibres;

(f) be elasticated at the cuffs, ankles and on the hoods of overalls and designed to ensure a close fit at the wrists, ankles, face and neck;

(g) not have pockets or other attachments which could attract and trap asbestos dust; and

(h) be easily decontaminated or disposable.

230 Where disposable overalls are used, these should be of a suitable standard.

14

Guidance

14

231 Disposable overalls which are Type 5 (under BS EN ISO 13982-1[21]), should be suitable.

232 Overall head coverings should be close-fitting and cover the parts of the head and neck not covered by the face piece of the respirator, and should be connected to the main overall. The head straps of RPE should be worn under the head covering. Wellington boots are preferable to any other form of footwear because they are easier to clean. Lace-up footwear will trap asbestos fibres between the laces and should not be worn.

233 Risks other than those created by potential exposure to asbestos should not be overlooked, for example where methods involve using equipment with naked flames the protective clothing should not be flammable.

ACOP

14

Removal of contaminated protective clothing

234 Protective clothing should be removed before taking off RPE. Protective clothing should also be removed before leaving the work area for any reason, including for meal breaks, for other breaks and at the end of the shift. Protective clothing should be vacuum-cleaned before removal using a Class 'H' (BS EN 60335) vacuum cleaner fitted with suitable attachments. If the clothing is to be reused (eg in licensed work), it should be placed in a storage area specifically provided for that purpose (eg in the airlock). If it is not to be reused, it should be placed in a suitable waste bag. If the clothing is to be removed from the premises for cleaning or disposal it should be sealed in a labelled, dust-tight bag.

235 If an enclosure is being used, and the main hygiene facilities are connected to the enclosure, then, following preliminary decontamination in the airlock, protective clothing, including footwear, should be taken off either in a connecting airlock or in the dirty end of the hygiene facility. If the main hygiene facilities are not connected to the enclosure, employers will need to provide additional overalls (transit overalls of a different colour to those worn inside the enclosure) for employees to wear after preliminary decontamination has taken place in the airlock to allow transfer to final decontamination at the main hygiene facilities. In addition, separate transit footwear should be provided for use between the airlock and the main hygiene facilities.

Cleaning, maintenance and storage of protective clothing

236 Non-disposable protective clothing and towels must be effectively washed after every shift. If the employer does not have the facilities and expertise for laundering asbestos-contaminated clothing, it must be sent to a specialist laundry.

237 Where disposable overalls are used they should be treated as asbestos waste and properly disposed of after every shift. Disposal after single use may not be necessary for overalls used for occasional sampling where there is a low risk of contamination.

238 Following work in enclosures, non-disposable clothing and towels for washing should be collected from the airlock and hygiene facility as soon as it has been discarded.

239 Asbestos-contaminated clothing for dispatch to a laundry should first be placed in dust-tight bags which are soluble in hot water and can be loaded, unopened, into a washing machine. These inner bags should then be placed inside a second bag which is labelled and which is strong enough to remain dust-tight during transport and handling. Dripping-wet overalls and other types of PPE

ACOP

should not be put into soluble bags as they may cause the bags to partially dissolve during transport, which could result in a release of dust when the outer bags are removed. Employers must make sure at the end of the working period that the bagged contaminated protective clothing is:

(a) placed in a specific storage area; or

(b) disposed of as asbestos waste (especially disposable overalls which should be disposed of after every shift); or

(c) prepared for dispatch to a laundry.

240 Asbestos surveyors and those sampling materials for asbestos who undertake occasional sampling as referred to in paragraph 237 should use their judgement to determine whether or not their overalls may have been contaminated and should be disposed of.

241 Contaminated protective clothing or materials, including contaminated towels, must never be taken home. Contaminated towels should be effectively washed after every shift or disposed of as contaminated waste.

242 Where the contaminated clothing is cleaned on the premises, or by a specialist laundry the washer and drier used must be dedicated for this use to prevent the spread of asbestos to other items of laundry. The room containing the washer and drier should have its own local exhaust ventilation, preferably an air mover fitted with HEPA filtration. The employee loading the washer should be wearing suitable RPE for protection. The air from the drier should be discharged to external atmosphere and on no account to an occupied workroom. Separate cycles should be used for heavily and lightly contaminated items.

243 The waste water from the washer should be filtered before going to drain. The filter should be treated as contaminated asbestos waste and when replaced
14 disposed of accordingly.

Regulation 15

Arrangements to deal with accidents, incidents and emergencies

Regulation

(1) Subject to regulation 3(2) and to paragraph (3) of this regulation, and without prejudice to the relevant provisions of the Management of Health and Safety at Work Regulations 1999,[a] *in order to protect the health of his employees from an accident, incident or emergency related to the use of asbestos in a work process or to the removal or repair of asbestos-containing materials at the workplace, the employer shall ensure that –*

(a) procedures, including the provision of relevant safety drills (which shall be tested at regular intervals), have been prepared which can be put into effect when such an event occurs;

15

(a) S.I. 1999/3242, as amended by S.I. 2003/2457 and S.I. 2006/438.

Regulation

15

(b) information on emergency arrangements, including –

(i) details of relevant work hazards and hazard identification arrangements, and

(ii) specific hazards likely to arise at the time of an accident, incident or emergency,

is available; and

(c) suitable warning and other communication systems are established to enable an appropriate response, including remedial actions and rescue operations, to be made immediately when such an event occurs.

(2) The employer shall ensure that information on the procedure and systems required by paragraph (1)(a) and (c) and the information required by paragraph (1)(b) is –

(a) made available to the relevant accident and emergency services to enable those services, whether internal or external to the workplace, to prepare their own response procedures and precautionary measures; and

(b) displayed at the workplace, if this is appropriate.

(3) Paragraph (1) shall not apply where –

(a) the results of the risk assessment show that, because of the quantity of asbestos present at the workplace, there is only a slight risk to the health of employees; and

(b) the measures taken by the employer to comply with the duty under regulation 11(1) are sufficient to control that risk.

(4) In the event of an accident, incident or emergency related to the unplanned release of asbestos at the workplace, the employer shall ensure that –

(a) immediate steps are taken to –

(i) mitigate the effects of the event,

(ii) restore the situation to normal, and

(iii) inform any person who may be affected; and

(b) only those persons who are responsible for the carrying out of repairs and other necessary work are permitted in the affected area and they are provided with –

(i) appropriate respiratory protective equipment and protective clothing, and

(ii) any necessary specialised safety equipment and plant,

which shall be used until the situation is restored to normal.

ACOP

15

Accidents, incidents and emergencies

244 Employers of people removing or repairing ACMs must have prepared procedures which can be put into effect should an accident, incident or emergency occur which could put people at risk because of the presence of asbestos unless, because of the quantity or the condition of the asbestos present at the workplace, there is only a slight risk to the health of employees.

Guidance

15

245 Such events may include:

(a) an employee collapsing or suffering a serious accident within an active stripping enclosure;

(b) emergency evacuation of the building, eg due to fire; or

(c) an uncontrolled release of asbestos, eg loss of containment in an active stripping enclosure, accidental spillage from waste bags, or damage to wrapped plant/pipework, which is being moved along the waste route.

ACOP

15

246 Sufficient information and instruction should be made available to the emergency services (eg fire and rescue and paramedics) so that when they are attending a relevant incident they can properly protect themselves against the risks from the asbestos.

Uncontrolled releases

247 In any circumstance where there is an accidental uncontrolled release of asbestos into the workplace then measures, including emergency procedures, should be in place to limit exposure and the risks to health. Such procedures should include means to raise the alarm and procedures for evacuation, which should be tested and practised at regular intervals. The cause of the uncontrolled release should be identified, and adequate control regained as soon as possible.

248 Any people in the work area affected who are not wearing PPE including RPE must leave that area. Where such people have been contaminated with dust or debris then arrangements should be made to decontaminate those affected. Any clothing or PPE should be decontaminated or disposed of as contaminated waste.

249 The contaminated area should be thoroughly cleaned of visible debris or dust that may have become contaminated by asbestos fibres using a suitable Class 'H' vacuum cleaner (BS EN 60335[9]). Employees doing this work must wear appropriate PPE, including RPE. Air sampling should then be carried out to confirm that the remedial measures taken have been effective.

250 It is essential for supervisors or managers to make a careful check to ensure the work has been properly carried out. Even if the work was non-licensable, a licensed contractor and analyst should be employed to thoroughly check and clean the area respectively if contamination is severe.

251 Only those people who are essential for carrying out repairs and other necessary cleaning and maintenance work must be allowed into the affected area (other than emergency services). For any employees who were not wearing adequate RPE or have been potentially exposed to asbestos fibres in an incident, a note that the exposure has occurred should be added to the employee's health record or to the employee's personal record if they do not have a health record. A copy of the note must be given to the employee with instructions that it should be kept indefinitely.

Regulation 16

Duty to prevent or reduce the spread of asbestos

Regulation 16

Every employer shall prevent or, where this is not reasonably practicable, reduce to the lowest level reasonably practicable the spread of asbestos from any place where work under his control is carried out.

ACOP

252 Any plant or equipment which has been contaminated with asbestos should be thoroughly decontaminated before it is moved for use or for disposal.

253 Any equipment used for the removal of asbestos which by its nature cannot be thoroughly decontaminated should be cleaned so far as is reasonably practicable and then suitably wrapped or otherwise sealed to prevent release of asbestos fibres. The outer surface of the wrapping or seal should then be thoroughly decontaminated before the plant or equipment is moved for use in other premises or for disposal. Such plant or equipment should not be unwrapped or unsealed other than in a manner which complies with the duties to prevent or reduce exposure, and to prevent or reduce the spread of asbestos.

16

254 The basic decontamination procedures described in paragraph 269-270 must be followed every time a person leaves the work area.

Guidance

16

255 Asbestos materials should never be left loose or in a state where they can be trampled, tracked over by plant and machinery or otherwise spread. All asbestos-contaminated waste should be removed at regular intervals in appropriate waste containers.

ACOP

Enclosures for non-licensable work

256 The expectation is that an enclosure will normally be required unless it is not reasonably practicable. Where there is a risk of significant contamination, the work area should be enclosed. For example, a full enclosure will be expected where there is large-scale work, eg textured decorative coating removal. A 'mini-enclosure' may be appropriate where the work is minor. Enclosures will not normally be required for work with asbestos cement.

Enclosures for licensable work

257 It should be assumed that for most of the work which requires a licence, which is not external/remote, a full enclosure will normally be required.

258 Enclosures should be designed and constructed so as to be able to fulfil their purpose in the particular circumstances. Employers should, as far as is reasonably practicable, make sure that the work area is completely enclosed to contain any asbestos debris and airborne asbestos fibres, either by erecting a purpose-made enclosure or by sealing the whole or part of the area where the work is to be carried out. Where the structure of a building forms part of the enclosure then particular attention should be paid to the effective sealing of areas such as windows, doors, vents and grilles and apertures through which pipes and other services/facilities pass which may allow air to escape and also to any surfaces which may not be readily cleanable. Heating, ventilation and air-conditioning systems should be turned off and sealed. The ends of any scaffold tubes used must be sealed.

16

Guidance

16

259 Openings in the enclosure for entry and exit ('airlocks' and 'bag locks') should be designed to prevent the escape of asbestos when people or waste bags pass through them and to permit the preliminary decontamination referred to in paragraphs 269-270. The airlocks should not reduce the effectiveness of the air

Guidance 16

extraction equipment. Air extraction units should have HEPA filtration and be of sufficient capacity to maintain a reduced air pressure within the enclosure to a level that is below that outside the enclosure and to ensure at least eight air changes per hour. The enclosure would then be at 'negative pressure'. The extraction unit(s) should be positioned at suitable places to ensure, wherever possible, that there are no 'dead spaces' (ie where there is little or no air movement) within the work area. Wherever reasonably practicable the filtered air from the enclosure should be discharged to the atmosphere external to any building. The discharge should, as far as is reasonably practicable be such that the discharged air is not likely to enter any building or other occupied space.

260 Information on the design and instructions for the installation and use of negative pressure units can be found in PAS 60-2:2004.[20]

ACOP 16

261 The enclosure, including airlocks, should be fitted wherever practicable with a sufficient number of suitable viewing panels. The panels should allow as much of the enclosed work area to be viewed from the outside as is practicable and they be should be kept clean and unobscured.

Guidance 16

262 As an alternative, or where viewing panels are impractical, eg in basements or upper floors or do not cover all areas, effective closed-circuit cameras or computer webcam systems should be used. Detailed guidance on the design and construction of enclosures can be found in *The licensed contractors' guide*.[13]

ACOP 16

263 The enclosure should normally be designed and constructed so that asbestos materials are not disturbed until the enclosure is complete and placed under negative pressure. In circumstances where the area that is to be enclosed is contaminated with asbestos debris which will be disturbed by the actual work to enclose the area, then the following action should be taken:

(a) As much of the area as practicable should be enclosed (and placed under negative pressure), taking care not to disturb the asbestos debris.

(b) The debris must then be cleaned up using methods to minimise fibre release and the enclosure completed. Where dust and debris must be cleaned up before the enclosure can be built, methods of minimising fibre release should be specified and adopted.

(c) Suitable PPE, including RPE, should be worn during pre-cleaning and for all work which disturbs or potentially disturbs asbestos during the building of enclosures.

264 Where scaffolding forms part of the enclosure or is liable to disturb asbestos while it is being erected then the employer of the scaffolders may require a licence and must take precautions under these Regulations. Any scaffolders who could be exposed to asbestos should be adequately trained as asbestos workers.

265 Before starting any work within the completed enclosure, its integrity should be checked by smoke testing. The filtered air extraction equipment should also be tested to ensure that it is achieving the required negative pressure and air changes.

Location of facilities and use of airlocks for licensable work

266 Where reasonably practicable, hygiene facilities should be connected directly to the enclosure airlock system. However, where this cannot be done, they should be located as close as is practicable, and procedures for preliminary decontamination and transiting should be drawn up and followed. The airlocks

ACOP

need to be of sufficient size (1 m x 1 m x 2 m minimum where space permits) to allow storage of equipment (eg vacuum cleaner, footbath with brush, separate bucket of water and sponge for wiping RPE and overalls) and to allow proper preliminary decontamination to take place. There should be weighted flaps on each of the airlocks, located on the enclosure side.

267 Where reasonably practicable, the 'transit route' should avoid occupied areas or, if that is not possible, the work should be carried out when the required transit areas are not occupied. If this cannot be done, then more rigorous decontamination of personnel will be required prior to transiting and more stringent arrangements may be required to deal with any incidents which may expose persons to asbestos.

16

Guidance 16

268 Detailed guidance on airlocks, including their design and construction, can be found in *The licensed contractors' guide*.[13]

ACOP

Preliminary decontamination procedures for licensable work

269 Employers must have in place clear procedures for exiting the enclosure and removing waste to prevent the spread of asbestos (regulation 16) and the subsequent potential risk of exposing others (regulation 11). The bulk of employee contamination should be removed during the preliminary decontamination procedure within the enclosure and airlocks, with only the residues being removed in the showers of the main hygiene facilities.

270 In addition to the main hygiene facilities, vacuum-cleaning equipment should be provided, which should be Class 'H' (BS EN 60335[9]) fitted with suitable tools, and be preferably located within the enclosure immediately next to the airlocks. Employees, using a 'buddy' system, should use the equipment to clean their protective clothing as thoroughly as possible whenever they leave the enclosure or work area. In transiting situations, in the inner stage of the airlock, footwear should be washed using a brush. Respirators (still worn and with the motor still running if a powered-assisted model is worn) should be wiped with wet cloths or sponges, using separate washing facilities to those provided for the footwear. All cloths, brushes and sponges should then be treated as contaminated waste. In the middle stage of the airlock, work overalls and boots should be removed. Transit overalls and footwear should be put on in the final compartment (outer stage) of the airlock.

Protective clothing for licensable work

271 Contaminated clothing should not be taken into the shower area or into the clean end of the hygiene facility.

Removal of waste during licensable work

272 Where practicable, waste bags should be removed from the enclosure via a separate bag lock. The bags should be decontaminated, eg thoroughly vacuumed and/or wet-wiped before being passed into the next compartment of the bag lock where the bags are put into second outer bags. The bags are then passed to the outside or to an additional storage compartment before being passed to the outside. Under no circumstances should people exit the enclosure via the bag exit. Where it is not practicable to have a separate bag lock system the bag lock should be constructed off the inner or middle stage of the three-stage airlock which provides the entry/exit system for people. Under no circumstances should waste bags be taken through the main hygiene facilities.

16

ACOP

273 All exits, whether for people or waste bags, should be designed and constructed to prevent or, if that is not possible, minimise the escape of airborne fibres and to allow 'negative pressure' equipment to operate effectively.

274 All waste should be transported between the enclosure and the skip or removal vehicle using the route likely to be the safest during normal transit and in the event of accidental release of asbestos.

Final decontamination procedure within main facilities for licensable work

275 RPE should not be removed until the wearer is in the shower and the respirator has been thoroughly washed. The exception to this is where transit arrangements are in operation and the nature of the site is such that to wear the equipment in transit would be dangerous for the wearer. This must be justified in the risk assessment. In such instances it is important that a suitable disposable respirator or half-mask respirator fitted with a particle filter is worn between the enclosure and the hygiene facilities. The exterior of the RPE should be thoroughly wiped clean before removal prior to transiting.

276 Parts of RPE which have been thoroughly cleaned in the hygiene facilities can be taken out through the 'clean' area. Contaminated equipment including any towels that have been taken into the shower or 'dirty end' will need to be put into a sealable container prior to being taken out through the 'dirty' area. Disposable towels should be treated as asbestos waste.

277 Once the removal of asbestos has started and until the area has been thoroughly cleaned ready for inspection, anyone (including analysts and supervisors) leaving the enclosure or working area should carry out preliminary and final decontamination (ie by properly using the main hygiene facilities including the shower). The only exceptions to this are where there is an acute risk to workers' health or safety due to a medical emergency in the enclosure.

16

Guidance

278 Adequate arrangements should be made to ensure that female workers can undergo decontamination procedures separately from males, and vice versa.

279 Analyst decontamination procedures during the 4-stage clearance will depend on whether the person has, or may have become, contaminated. The potential for contamination will reflect the conditions inside the enclosure and the activity undertaken while in there. The analyst should always carry out preliminary decontamination on leaving the enclosure including in situations where there is no obvious visible contamination. Full decontamination (preliminary and final decontamination) would be expected in situations where the analyst may have become contaminated with asbestos or material suspected of containing asbestos. An example would be after a failed visual inspection where RPE or PPE including footwear could have become contaminated by the intrusive nature of the inspection, eg where the analyst has had to crawl through confined areas such as an undercroft or ceiling space, or any other such area where significant contamination was possible.

280 Analysts should wear appropriate protective clothing for the 4-stage clearance procedure. As there is always the possibility that the analyst may become contaminated (and therefore need to undergo full decontamination), all PPE including undergarments, should be disposable or be able to be cleaned. Arrangements should be in place to ensure that at least one set of clothes, shoes and towel are always available in the clean end of the hygiene facility. Under no circumstances should an analyst take or wear contaminated personal clothing home.

16

ACOP

Licensable work in open sites

281 If it is not reasonably practicable to enclose the work area (eg on exposed or remote sites) then the work area should be marked by suitable warning notices and by physical barriers, which should be placed appropriately. Employers must assess the risks to workers and others nearby and, if necessary and as far as is reasonably practicable, the work should be done when other workers or members of the public will not be in the vicinity. Where it is not reasonably practicable to build a full enclosure, the spread of fibres should be prevented by other containment and dust-suppression techniques.

16

Guidance

282 For example, techniques such as wrap-and-cut where enclosures are used at the cut points, or glovebags combined with controlled wet stripping may be more appropriate for this. In addition, partial enclosures can be used for asbestos soffit removal. Guidance on the choice of asbestos stripping techniques is given in *The licensed contractors' guide.*[13]

16

ACOP

283 Where enclosures are not used, particular attention should be given to the risk assessment to establish what will be required to ensure that, as far as is reasonably practicable, the spread of asbestos is prevented and the work area is thoroughly cleaned.

16

Regulation 17

Cleanliness of premises and plant

Regulation

Every employer who undertakes work which exposes or is liable to expose his employees to asbestos shall ensure that –

(a) the premises, or those parts of the premises where that work is carried out, and the plant used in connection with that work are kept in a clean state; and

(b) where such work has been completed, the premises, or those parts of the premises where the work was carried out, are thoroughly cleaned.

17

ACOP

Cleanliness of premises and plant

284 When work with asbestos comes to an end, the work area should be thoroughly cleaned before being handed over for reoccupation or for demolition. All visible traces of asbestos dust and debris should be removed and a thorough visual inspection carried out. Where the work is licensable then the 4-stage clearance procedure (which includes air sampling) should be carried out and a certificate of reoccupation issued. Where licensed work is performed out of doors (eg soffit removal), then clearance air sampling will not be required. In this situation, the certificate of reoccupation should still be completed but without stage 3 (air monitoring). Clearance air sampling will not be required for non-licensed work.

17

Guidance

285 More information on clearance procedures for non-licensed work is given in the *Asbestos essentials task manual.*[10]

17

ACOP

286 To aid the process of cleaning and to prevent the spread of asbestos, employers must choose work methods and equipment to prevent the build-up of asbestos waste on floors and surfaces in the working area. Wherever practicable, waste should be transferred direct into waste bags as workers remove the asbestos materials. Employers must make sure that any asbestos dust and debris is cleaned

17

ACOP 17

up and removed regularly to prevent it accumulating (and drying out where wet removal techniques have been used), and at least at the end of each shift.

Guidance 17

287 Surveyors and others taking samples are responsible for cleaning up any material which they have disturbed, but are not responsible for cleaning up pre-existing dust and debris.

ACOP

288 Procedures will need to take account of the necessity for cleaning following an accidental and uncontrolled release of asbestos.

Further measures to keep premises and plant clean during licensable work

289 Procedures will need to be set up for cleaning:

(a) working areas including transit and waste routes;

(b) plant and equipment; and

(c) hygiene facilities.

290 Dustless methods of cleaning should be used including, wherever practicable, a Class 'H' vacuum cleaner (BS EN 60335[9]) with appropriate tools. Procedures for cleaning should make clear:

(a) the items and areas to be cleaned;

(b) how often they need to be cleaned;

(c) the cleaning methods, which should not create dust; and

(d) any special precautions which need to be taken during cleaning, including the low-dust technique to be used, and the measures to be taken to reduce the spread of dust.

291 Dry manual brushing, or sweeping or compressed air must not be used to remove asbestos dust.

Site clearance certification for reoccupation

Site clearance duties and roles

292 The employer of the people carrying out work with the asbestos or ACMs has duties:

(a) to ensure other people are not exposed to asbestos;

(b) to prevent the spread of asbestos; and

(c) to ensure that the premises or parts of premises where work with asbestos has taken place are thoroughly cleaned.

293 Compliance with these duties is aided by:

17

(a) pre-cleaning where necessary;

ACOP

(b) choosing methods which reduce the amount of airborne asbestos to the lowest level reasonably practicable;

(c) controlling the waste produced;

(d) using enclosures to prevent spread;

(e) the thorough cleaning of the work area and areas which may have become contaminated;

(f) visual inspection of the work area and areas which may have become contaminated;

(g) obtaining a clearance certificate for reoccupation of the area and a separate clearance certificate for the hygiene facility.

The process for site clearance certification for reoccupation

294 Once removal of the asbestos has been completed, the premises must be assessed to determine whether they are thoroughly clean and hence fit to be returned to the client. It is important that this includes the premises, any plant or equipment or parts of the premises where work with asbestos has taken place and the surrounding areas which may have been contaminated. The areas requiring assessment for site clearance certification for reoccupation include:

(a) the enclosed area including airlocks or the delineated work area where an enclosure has not been used;

(b) the immediate surrounding area (for enclosures this will include the outside of walls and underneath polythene floors; for delineated areas this will include surfaces nearby either where asbestos may have been spread or where the pre-cleaning was not done properly);

(c) the transit route if one has been used; and

(d) the waste route and area around the waste skip.

295 Those employing an organisation to carry out air testing as part of certification for reoccupation must ensure that the organisation is accredited to meet the relevant criteria in ISO 17020[22] and ISO 17025.[23] From 1 April 2007, anyone engaged to carry out site clearance certification for reoccupation must demonstrate that they conform with specified requirements in ISO 17020 and ISO 17025 through accreditation with a recognised accreditation body. The process is intended to allow flexibility, but in practice it is likely to be the same person or organisation who carries out each stage. This will aid continuity and consistency, and will avoid problems with interfaces at each stage of the process. The organisation should have the necessary independence to act completely impartially. If the analyst is contracted by the client then a copy of the clearance certificate should be made available to the asbestos removal company.

17

Guidance

296 Although not a legal requirement, it is desirable that the analyst is employed by the building owner or occupier for site clearance certification. This arrangement helps avoid any conflict of interest (perceived or real) that may arise should the analyst be employed by the removal contractor. It also enables an independent party to be involved in resolving any problems that arise during the clearance process. In addition, it has a practical advantage in that all results and certificates of reoccupation can also be issued directly to the person who has responsibility for the premises as well as to the contractor.

17

ACOP

297 Site clearance certification for reoccupation should only be carried out when work has been completed and the employer of those who have carried out the work has ensured that the areas requiring clearance assessment have been thoroughly cleaned and allowed to dry. To do this employers should follow the guidance given in paragraphs 311-312 on checking site condition, job completeness and carrying out a thorough visual inspection. Site clearance certification for reoccupation should normally be carried out in four successive stages, with the next stage only being commenced when the previous one has been completed.

17

Guidance

298 However, more complex jobs, eg where multi-enclosure clearance is required or where scaffolding is remaining on site, may involve an alteration to the clearance sequence. For example, common areas (such as the transit route and the area around the skip) on multi-enclosure sites would only be cleaned at final completion of the job.

17

ACOP

299 In situations where more clearance stages are required, this should be taken account of in the plan of work.

300 The four stages of site clearance certification for reoccupation are:

(a) stage one – preliminary check of site condition and job completeness;

(b) stage two – a thorough visual inspection inside the enclosure/work area;

(c) stage three – clearance air monitoring;

(d) stage four – final assessment post enclosure/work area dismantling.

301 Where practical the areas to be assessed should be dry and therefore sealants (such as PVA) should not be used prior to any visual inspections or disturbed air tests. Where it is not practical for the area to be dry (eg where there is natural water ingress) this fact should be recorded by the contractor before the site clearance process commences.

302 Occasionally some surfaces or materials, eg concrete, require sealing before the disturbed air test because they produce quantities of non-asbestos dust which would lead to an apparent failure in the air test. The use of sealants in this circumstance should only be done under the direction of the person carrying out the air test and the fact recorded by the contractor before the clearance process commences.

303 In some circumstances the floor of an enclosure may be covered with a 'sacrificial' layer of suitable floor material to prevent damage to the polythene underneath it, reduce the risk of slips and allow safe use of access equipment. Dust or debris may have penetrated between the sacrificial layer and the polythene, and therefore it will become necessary to take up the covering before site clearance certification.

Stage one: Preliminary check of site condition and job completeness

304 The scope of clearance should be established. The plan of work kept at the site should be checked and the extent of the clearance being sought agreed between the analyst and the contractor. The scope of the clearance should be recorded (eg on a diagram). A note should be made of any remaining asbestos which was outside the scope of the work.

17

ACOP

Stage two: thorough visual inspection

305 The work area, enclosure, hygiene facilities, and controls should be intact, operating and clean, with all ACMs included in the scope of the work and non-essential equipment decontaminated and removed. The hygiene facilities should remain operable until a certificate of reoccupation has been issued. The work area, surrounding area, transit route, waste route together with the area around the waste disposal storage and all sections of the hygiene facility must be free of obvious asbestos-containing waste and debris of any kind. If a viewing panel is fitted, this should be looked through so that a preliminary check can be made of the inside of the enclosure to see whether it contains any waste and debris. The result of these pre-inspections should be recorded.

306 A thorough visual inspection should then be carried out to make sure that all visible traces of asbestos and other dust and debris have been removed, as far as is reasonably practicable, from the enclosure (including airlocks) or work area. It is important to refer to the plan of work to check that all the asbestos that was due to be removed has been removed. To be thorough this visual inspection should consist of the following three checks:

(a) the completeness of the removal of the ACM from the underlying surfaces;

(b) the presence of any visible asbestos debris left inside the enclosure and airlocks or work area where there is no enclosure; and

17 (c) the presence of fine settled dust.

Guidance 17

307 Suitable facilities should be made available to enable the inspection to be properly carried out, eg access platforms so that higher levels of the enclosure can be inspected.

ACOP

Stage three: clearance air monitoring

308 Following the successful completion of the thorough visual inspection, and before the enclosure is dismantled or the work area handed back to the client, air monitoring should be carried out to check that the concentration of airborne fibres remaining in areas affected by the work is as low as is reasonably practicable. For enclosures this is carried out with the enclosure intact and dry, but with the negative pressure unit switched off and the pre-filter capped and sealed.

309 The monitoring should be accompanied by activities which raise dust from the surfaces at least to a level consistent with normal use of the area and possible future work activities. The type of disturbance method and the length of time it is carried out for should be recorded. As many areas will subsequently be subjected to normal cleaning activities, air disturbance tests should be carried out using a brush to raise potential dust. Any person carrying out air disturbance must wear appropriate PPE/RPE. For work areas without enclosures, reassurance or background air testing is more appropriate than a disturbed air test. In most cases it will be reasonably practicable to clean the working area thoroughly enough for the airborne fibre concentration in the enclosure/work area, after final cleaning, to be less than 0.01 f/cm^3. If measurements of 0.01 f/cm^3 or more are found, an investigation will need to be carried out to find out the cause. If it is found that the enclosure or work area has not been cleaned properly then it must be re-cleaned, visually inspected and re-monitored. The threshold of less than 0.01 f/cm^3 should be taken only as a transient indication of site cleanliness, in conjunction
17 with visual inspection, and not as an acceptable permanent environmental level.

Guidance 17

310 Details of the method to use when undertaking clearance air sampling can be found in *The analysts' guide*.

ACOP

Stage four: final assessment post enclosure/work area dismantling

311 Once the enclosure or work area has passed the visual inspection and clearance air monitoring the enclosure or work area can be dismantled. A Class H vacuum cleaner (BS EN 60335[9]) and suitable PPE, including RPE, should be kept available during dismantling so that any small amounts of asbestos debris which have become lodged behind the fabric of the enclosure or within folds in the polythene sheeting or on the floor underneath can be removed. Once the enclosure or work area has been dismantled, the area should be visually inspected again by a competent person from the accredited organisation, as detailed in paragraph 306, to ensure that all debris has been removed.

312 Where there is evidence of dust and debris being released during dismantling of the enclosure, and this cannot be easily removed by vacuum, the site should be re-enclosed, re-cleaned, the visual inspection repeated and a disturbed air test carried out to make sure that the airborne asbestos fibre concentration is as low as is reasonably practicable, and in any case below the clearance indicator.

Clearance certification

313 Taking into account the results of each of the four stages of the clearance process, a certificate of reoccupation should be issued when the area concerned is deemed to be clean and cleared and suitable for return to the client. The certificate should include details of the site address, the dates of the work and a brief description, the name of the contractor, details of the clearance action that was undertaken under each stage and the specific areas and items checked, the results of each stage, and the signature of the person completing each stage.

314 For premises permanently set aside for the testing and maintenance of plant and equipment contaminated with asbestos, the measures set out in paragraph 286 should be followed in order to keep the area clean. When such an area is to be used for non-asbestos work then the area will need to be thoroughly cleaned, the clearance process carried out and a site certificate of reoccupation issued beforehand.

Clearance testing of hygiene facilities

315 Once the certificate of reoccupation has been issued for the work area, a clearance test should be carried out on the hygiene facility before it is removed from the site. The facility should be visually inspected and air tested. There should be a thorough visual inspection of all sections (ie clean end, showers and dirty end). The unit, including the shower, should be dry before the inspection takes place. On successful completion of the visual examination, a disturbed air test should be performed in the shower and dirty end. Clearance testing should be performed by a competent person. A separate clearance certificate should be issued for the hygiene facility. A copy of the most recent clearance certificate should be kept with the facility.

Duties of those issuing clearance certificates

316 The person who issues the site clearance certificate for reoccupation or the clearance certificate for the hygiene facility does not have direct duties under the Asbestos Regulations. However, people issuing these certificates should follow this
17 guidance to comply with their duty under section 3 of the HSW Act to protect the

ACOP

17

health of people other than their employees. They should also consider the provision in section 36 of the HSW Act[3] which may become operative if they cause other people having duties under these Regulations to fail in those duties.

Regulation 18

Designated areas

Regulation

(1) Every employer shall ensure that any area in which work under his control is carried out is designated as –

(a) an asbestos area, subject to regulation 3(2), where any employee would be liable to be exposed to asbestos in that area;

(b) a respirator zone where the concentration of asbestos fibres in the air in that area would exceed or would be liable to exceed the control limit.

(2) Asbestos areas and respirator zones shall be clearly and separately demarcated and identified by notices indicating –

(a) that the area is an asbestos area or a respirator zone or both, as the case may be; and

(b) in the case of a respirator zone, that the exposure of an employee who enters it is liable to exceed the control limit and that respiratory protective equipment must be worn.

(3) The employer shall not permit any employee, other than an employee who by reason of his work is required to be in an area designated as an asbestos area or a respirator zone, to enter or remain in any such area and only employees who are so permitted shall enter or remain in any such area.

(4) Every employer shall ensure that only competent employees shall –

(a) enter a respirator zone; and

(b) supervise any employees who enter a respirator zone,

and for the purposes of this paragraph a competent employee means an employee who has received adequate information, instruction and training.

(5) Every employer shall ensure that –

(a) his employees do not eat, drink or smoke in an area designated as an asbestos area or a respirator zone; and

(b) arrangements are made for such employees to eat or drink in some other place.

18

ACOP

Designated areas for asbestos work

317 All areas where asbestos work is being undertaken should be demarcated and identified by suitable warning notices as asbestos areas, subject to the exemptions provided in regulation 3(2). Where it is not necessary to demark an asbestos area (due to the exemptions in regulation 3(2)) or respirator zone (because the control limit will not be exceeded, or is not liable to be exceeded), RPE should still be worn if it is reasonably practicable to do so.

18

Guidance 18

318 In many cases the work area will be contained within an enclosure. Warning notices should be placed on the airlock and enclosure walls. In the absence of an enclosure, the boundaries to the work area should be clearly established at a suitable distance from the work by physical obstructions such as ropes or barriers. Warning notices should be placed at suitable locations around the work area and at the entrance.

ACOP 18

319 Any area where an employee may be exposed to asbestos to a level which may exceed a control limit must be designated as a respirator zone. Respirator zones, whether enclosed or not, must be demarcated and identified by suitable warning notices including notices that RPE must be worn.

Guidance 18

320 Respirator zone warning notices should be placed at similar locations to those for the asbestos area.

321 In most situations the asbestos area and respirator zones will be the same. However, in certain circumstances where work is being carried out within a respirator zone, it may be necessary to mark out an asbestos area outside the respirator zone to keep people away from the area where ancillary work activities are being carried out.

ACOP 18

322 Only employees who need to do so for their work can enter and remain in asbestos areas and respirator zones.

323 Only employees who are competent may enter respirator zones or supervise people working in respirator zones. To enter a respirator zone the employee must have received adequate information, instruction and training in accordance with regulation 10 and other relevant health and safety legislation. Persons undergoing training may enter a respirator zone provided they are under the direct supervision of a competent person.

Guidance 18

324 Other essential personnel such as analysts, tradesmen and emergency services may enter a respirator zone provided they have received adequate information, instruction and training and are wearing adequate RPE and PPE.

ACOP 18

325 Employers should ensure the provision of suitable facilities for employees to eat and drink outside the working area. For licensed work, the eating facilities should be located as close as is reasonably practicable to the hygiene facilities. No one should eat, drink or smoke in the enclosure or work area, in the hygiene facilities or in any areas which have been marked as asbestos areas or respirator zones.

326 Employers should also ensure that toilet facilities are provided, if they are not provided elsewhere on the site.

Guidance 18

327 Where hygiene facilities are not required, personnel should clean and decontaminate themselves whenever they leave an asbestos area or respirator zone.

Regulation 19

Air monitoring

Regulation

(1) Subject to paragraph (2), every employer shall monitor the exposure of his employees to asbestos by measurement of asbestos fibres present in the air –

(a) at regular intervals; and

(b) when a change occurs which may affect that exposure.

(2) Paragraph (1) shall not apply where –

(a) the exposure of an employee is not liable to exceed the control limit; or

(b) the employer is able to demonstrate by another method of evaluation that the requirements of regulation 11(1) and (5) have been complied with.

(3) The employer shall keep a suitable record of –

(a) monitoring carried out in accordance with paragraph (1); or

(b) where he decides that monitoring is not required because paragraph 2(b) applies, the reason for that decision.

(4) The record required by paragraph (3), or a suitable summary thereof, shall be kept –

(a) in a case where exposure is such that a health record is required to be kept under regulation 22 for at least 40 years; or

(b) in any other case, for at least 5 years,

from the date of the last entry made in it.

(5) In relation to the record required by paragraph (3), the employer shall –

(a) on reasonable notice being given, allow an employee access to his personal monitoring record;

(b) provide the Executive with copies of such monitoring records as the Executive may require; and

19 *(c) if he ceases to trade, notify the Executive forthwith in writing and make available to the Executive all monitoring records kept by him.*

ACOP

Air monitoring

328 Personal sampling/air monitoring is required to protect the health of employees by determining or checking the concentrations of airborne asbestos to which they are exposed, the adequacy of the controls and RPE and to establish employee exposure records. This should be done at regular intervals for a representative range of jobs and work methods. Where groups of workers are doing the same type of work in similar conditions, sampling can be carried out on a group basis. Individuals chosen for sampling within a group should be selected at random.

19

ACOP

329 Static sampling/air monitoring can be used to measure the concentration of asbestos fibres present in the work area and for background, leak and reassurance sampling, to check that control measures are effective. It is also used for clearance sampling to check that the enclosure has been properly cleaned after completion of the asbestos removal work.

330 Air monitoring should always be done when there are any doubts about the effectiveness of the measures taken to reduce the concentration of asbestos in air (eg that engineering controls are working as they should to their design specification and do not need repair), and, in particular, measures taken to reduce that concentration below the control limit or below a peak level measured over 10 minutes of 0.6 f/cm^3, as detailed in paragraph 161. Personal air monitoring will also be necessary to confirm that the RPE chosen will provide the appropriate degree of protection where the level of asbestos fibres in air exceeds, or is liable to exceed, the control limit or a peak level measured over 10 minutes of 0.6 f/cm^3.

331 Air monitoring will be appropriate unless:

(a) previous measurements have shown that the airborne concentrations are low and not likely to approach the control limit or a 10 minute peak of 0.6 f/cm^3;

(b) the work is such that it complies with regulation 3(2) and adequate information is available to enable the appropriate protective equipment to be provided; or

(c) the protective equipment provided is of such a standard that no foreseeable measurement could indicate a need for equipment of a higher standard.

19

Guidance

332 If the employer decides air monitoring is no longer necessary then he or she should list the values and sources of information about the likely concentrations of asbestos in air for the activities and tasks being carried out. This should include previous air monitoring data on the employees from the same or similar type of removal job or relevant published data (for instance similar activities with similar material being disturbed as given in *The licensed contractors' guide*[13]).

333 Analysis must be undertaken using the 1997 WHO recommended method.

334 Further information and guidance on the sampling strategy, the methods for sampling and analysis and the reporting of results of air monitoring can be found in *The analysts' guide*.[8]

19

ACOP

335 All records of air monitoring will need to state the employer's name and address, the site address where appropriate and the date of air monitoring, and should also include:

(a) the type of work being done and, where relevant, its exact location;

(b) the type of sample, eg personal, static, clearance etc;

(c) the location of any static sampler;

(d) the date and time of sampling, the sample duration and the flow rate;

(e) if a personal sample, the employee's name, task being performed and the category of RPE being worn;

19

(f) the length of time for which individuals are exposed;

ACOP

(g) the measured fibre concentration;

(h) the fibre type, if known; and

(i) the names and organisations of the sampler and analyst and the sampling and analysis method used.

336 Records of air monitoring or a suitable summary must be kept for 5 years except that, where employees are under medical surveillance, employers must keep the records or summary to supplement the health record for 40 years. Any summary of results will need to contain enough information about airborne fibre levels to allow individual average exposures for different types of work to be estimated as accurately as possible.

337 Employers should consult employees, safety representatives or representatives of employee safety when making arrangements for monitoring. On reasonable notice being given, the records or summary of the airborne fibre monitoring must be made available to employees.

19

Guidance 19

338 The results from all personal monitoring carried out by the licensed asbestos removal contractor during the period of the licence, should be collated and submitted to HSE as part of the licence renewal procedure.

Regulation 20

Standards for air testing and site clearance certification

Regulation

(1) In paragraph (4) "site clearance certificate for reoccupation" means a certificate issued to confirm that premises or parts of premises where work with asbestos has been carried out have been thoroughly cleaned upon completion of that work in accordance with regulation 17(b).

(2) Every employer who carries out any measurement of the concentration of asbestos fibres present in the air shall ensure that he meets criteria equivalent to those set out in the paragraphs of ISO 17025 which cover organisation, quality systems, control of records, personnel, accommodation and environmental conditions, test and calibration methods, method validation, equipment, handling of test and calibration items, and reporting results.

(3) Every employer who requests a person to carry out any measurement of the concentration of asbestos fibres present in the air shall ensure that that person is accredited by an appropriate body as competent to perform work in compliance with ISO 17025.

(4) Every employer who requests a person to assess whether premises or parts of premises where work with asbestos has been carried out have been thoroughly cleaned upon completion of that work and are suitable for reoccupation such that a site clearance certificate for reoccupation can be issued shall ensure that that person is accredited by an appropriate body as competent to perform work in compliance with the paragraphs of ISO 17020 and ISO 17025 which cover organisation, quality systems, control of records, personnel, accommodation and environmental conditions, test and calibration methods, method validation, equipment, handling of test and calibration items, and reporting results.

(5) Paragraphs (2) and (3) shall not apply to work carried out in a laboratory for the purposes only of research.

20

ACOP 20

339 Those engaged to carry out air measurements and employee exposure monitoring must demonstrate that they conform with specified requirements in ISO 17025 through accreditation with a recognised accreditation body.

340 Employers carrying out their own air measurements or employee exposure monitoring should make sure that employees carrying out this work receive similar standards of training, supervision and quality control to those required by ISO 17025.

341 From 6 April 2007, those engaged to carry out site clearance certification for reoccupation must demonstrate that they conform with specified requirements in ISO 17020 and ISO 17025 through accreditation with a recognised accreditation body.

Guidance 20

342 The United Kingdom Accreditation Service (UKAS) is currently the sole recognised accreditation body in Great Britain.

Regulation 21

Standards for analysis

Regulation 21

(1) Every employer who analyses a sample of any material to determine whether it contains asbestos shall ensure that he meets criteria equivalent to those set out in the paragraphs of ISO 17025 which cover organisation, quality systems, control of records, personnel, accommodation and environmental conditions, test and calibration methods, method validation, equipment, handling of test and calibration items, and reporting results.

(2) Every employer who requests a person to analyse a sample of any material taken to determine whether it contains asbestos shall ensure that that person is accredited by an appropriate body as competent to perform work in compliance with ISO 17025.

(3) Paragraphs (1) and (2) shall not apply to work carried out in a laboratory for the purposes only of research.

ACOP 21

343 Those engaged to analyse samples of material to determine whether or not they contain asbestos must demonstrate that they conform with ISO 17025 by accreditation with a recognised accreditation body.

344 Employers carrying out their own analysis of samples should make sure that employees carrying out this work receive similar standards of training, supervision and quality control to those required by ISO 17025.

Guidance 21

345 The United Kingdom Accreditation Service (UKAS) is currently the sole recognised accreditation body in Great Britain.

Regulation 22

Health records and medical surveillance

Regulation

(1) Subject to regulation 3(2), every employer shall ensure that –

(a) a health record, containing particulars approved by the Executive, relating to each of his employees who is exposed to asbestos is maintained; and

(b) that record or a copy thereof is kept available in a suitable form for at least 40 years from the date of the last entry made in it.

(2) Subject to regulation 3(2), every employer shall ensure that each of his employees who is exposed to asbestos is under adequate medical surveillance by a relevant doctor.

(3) The medical surveillance required by paragraph (2) shall include –

(a) a medical examination not more than 2 years before the beginning of such exposure; and

(b) periodic medical examinations at intervals of not more than 2 years or such shorter time as the relevant doctor may require while such exposure continues,

and each such medical examination shall include a specific examination of the chest.

(4) Where an employee has been examined in accordance with paragraph (3), the relevant doctor shall issue a certificate to the employer and employee stating –

(a) that the employee has been so examined; and

(b) the date of the examination,

and the employer shall keep that certificate or a copy thereof for at least 4 years from the date on which it was issued.

(5) An employee to whom this regulation applies shall, when required by his employer and at the cost of the employer, present himself during his working hours for such examination and tests as may be required for the purposes of paragraph (3) and shall furnish the relevant doctor with such information concerning his health as the relevant doctor may reasonably require.

(6) Where, for the purpose of carrying out his functions under these Regulations, a relevant doctor requires to inspect any record kept for the purposes of these Regulations, the employer shall permit him to do so.

(7) Where medical surveillance is carried out on the premises of the employer, the employer shall ensure that suitable facilities are made available for the purpose.

(8) The employer shall –

(a) on reasonable notice being given, allow an employee access to his personal health record;

22

Regulation

(b) *provide the Executive with copies of such personal health records as the Executive may require; and*

(c) *if he ceases to trade, notify the Executive forthwith in writing and make available to the Executive all personal health records kept by him.*

(9) Where, as a result of medical surveillance, an employee is found to have an identifiable disease or adverse health effect which is considered by a relevant doctor to be the result of exposure to asbestos at work the employer of that employee shall –

(a) *ensure that a suitable person informs the employee accordingly and provides the employee with information and advice regarding further medical surveillance;*

(b) *review the risk assessment;*

(c) *review any measure taken to comply with regulation 11 taking into account any advice given by a relevant doctor or by the Executive;*

(d) *consider assigning the employee to alternative work where there is no risk of further exposure to asbestos, taking into account any advice given by a relevant doctor; and*

(e) *provide for a review of the health of every other employee who has been similarly exposed, including a medical examination (which shall include a specific examination of the chest) where such an examination is recommended by a relevant doctor or by the Executive.*

22

ACOP

Health records and medical surveillance for licensable work

346 The employer must keep a health record for any employee who undertakes licensable work. The health record must be kept for 40 years in a safe place and should contain at least the following information:

(a) the individual's surname, forenames, sex, date of birth, permanent address, postcode and National Insurance number;

(b) a record of the types of work carried out with asbestos and, where relevant, its location, with start and end dates, with the average duration of exposure in hours per week, exposure levels and details of any RPE used;

(c) a record of any work with asbestos prior to this employment of which the employer has been informed; and

(d) dates of the medical examinations.

22

Guidance

347 The health record is not a medical confidential record. It contains demographic and job exposure information and the dates of any medical examinations as required by regulation 22. This health record must be kept for at least 40 years after the last entry made in it. Considering the long latency time of asbestos-related diseases employers should keep the health record for at least 40 years after last entry or at least until the employee would reach the age of 80 years, whichever ensures the longer retention time. The information the employee provides to the appointed doctor as part of the medical surveillance and any findings and recommendations from the appointed doctor will be documented in a medical record. The appointed doctor will keep this medically confidential record in line with professional arrangements.

22

ACOP

348 Anyone who undertakes licensable work must have been medically examined within the previous two years. Employers will need to obtain certificates of examination for any employees who state that they have been examined under these Regulations within the previous two years and keep them for four years from the date of issue. Employers should check with the previous employer or with the examining doctor that the certificates are genuine.

349 Medical examinations should take place during the employee's normal working hours and be paid for by the employer. Employees should co-operate with their employer regarding attendance for medical examinations.

350 Where an employee is diagnosed with a condition related to exposure to asbestos then the employer must review the health of all other current employees similarly exposed, as well as reviewing his assessments and methods of work.

351 If the examination reveals the presence of any potentially limiting health conditions then a decision should be reached on whether a general fitness assessment is required in addition to the asbestos medical examination.

22

Guidance

352 The certificate of examination can only state the date of the examination and that the employee has been examined as required by regulation 22. It does not certify that the person is fit to work with asbestos or fit to work in the conditions usually associated with licensed asbestos work, eg strenuous activities or removal in confined or restricted spaces. However, the medical assessment may bring information to light which the appointed doctor is obliged to advise the employee, eg the person has a condition such that they are not fit to work with asbestos. The appointed doctor will record the information in the medical file. Where the condition means that the person is clearly a danger to themselves and possibly others, the appointed doctor will consider informing the employer. Usually, however, the appointed doctor will seek the person's informed consent before disclosing any medical in confidence information to the employer. These ad hoc arrangements will lead to variation and inconsistencies and may not comply with fitness assessment requirements under other regulations (eg Management of Health and Safety at Work Regulations 1999, Confined Spaces Regulations 1997[24]). Therefore employers should consider formally arranging for a fitness-for-work examination to be carried out in addition to the asbestos medical. The examination may also be necessary as a consequence of the asbestos risk assessment.

22

Regulation 23

Washing and changing facilities

Regulation

(1) Every employer shall ensure that, for any of his employees who is exposed or liable to be exposed to asbestos, there be provided –

(a) adequate washing and changing facilities;

(b) where he is required to provide protective clothing, adequate facilities for the storage of –

(i) that protective clothing, and

(ii) personal clothing not worn during working hours; and

(c) where he is required to provide respiratory protective equipment, adequate facilities for the storage of that equipment.

23

Regulation

(2) The facilities provided under paragraph (1) for the storage of –

(a) personal protective clothing;

(b) personal clothing not worn during working hours; and

(c) respiratory protective equipment,

shall be separate from each other.

23

ACOP

353 The type and extent of washing and changing facilities provided should be determined by the type and amount of exposure as indicated by the risk assessment.

Hygiene facilities for licensable work

354 If work which requires a licence is carried out, suitable and sufficient hygiene facilities must be provided to enable those working with asbestos to be able to clean and decontaminate themselves, to help prevent the spread of asbestos and to reduce the risk of exposure of others. Suitable hygiene facilities (whether purpose-built on site or a transportable dedicated decontamination unit (DCU)), must be provided on site and fully operational before any work (including ancillary work) commences. Maintenance records for DCUs (or copies of them) should be kept on site. The hygiene facility should remain operational and not leave the site until the job is complete and the certificate of reoccupation has been issued.

355 The hygiene facilities will need to:

(a) have separate changing rooms for dirty, contaminated work clothing and for clean or personal clothing known as 'dirty' and 'clean' areas respectively. The showers should be located between the two changing rooms so that it is necessary to pass through them when going from one changing facility to the other. All doors between each room and those leading to the outside from the 'dirty end' should be self-closing and provide an airtight seal. The 'clean' and 'dirty' ends should be fitted with adequate seating and be of sufficient size for changing purposes;

(b) be designed so that they can be cleaned easily;

(c) be fitted with effective air extraction equipment which maintains a flow of air from the clean to the dirty areas. The extracted air must be discharged through a HEPA filter;

(d) be adequately heated, adequately lit (with suitable light switches at both the 'clean' and 'dirty' ends) and have suitable internal vents so that air can pass through the unit;

(e) be of sufficient size, including allowance for sufficient and separate storage for personal clothing and protective clothing and equipment in the 'clean' end and sufficient, suitable receptacles for contaminated clothing, towels, filters etc in the 'dirty' end and shower area;

(f) have showers with an adequate supply of clean running hot and cold or warm water, at a suitable pressure. Sufficient soap or gel, shampoo, nail brushes and individual dry towels must be provided for the asbestos workers and for any other person who may need to use the facilities for decontamination;

23

ACOP

(g) have all waste water filtered before it is discharged to the drains. All filters should be treated as asbestos waste;

(h) be adequately heated. Any gas heater mounted inside the unit must be of a room-sealed type; open-flued types must not be used. All gas appliances and fittings should be maintained in a safe condition by a competent person;

(i) have shower areas of sufficient size to allow thorough decontamination and to have means to support the power pack of a full face respirator while it is still required to be worn (the power pack support should be out of the direct line of the shower to avoid contact with water and prevent damage to the batteries);

(j) have a wall-mounted mirror in the clean end of the unit;

(k) have the electricity supply routed via a 30 mA residual current circuit-breaker fitted at the point of entry into the unit, and the unit must be effectively earthed when in use. The electrical fittings and installation must be suitable for use in the facility and maintained in a safe condition by a competent person;

(l) be sufficient for the number of people likely to need them; and

23 (m) be maintained in a safe condition and kept clean as far as is reasonably practicable.

Regulation 24 Storage, distribution and labelling of raw asbestos and asbestos waste

Regulation

(1) Every employer who undertakes work with asbestos shall ensure that raw asbestos or waste which contains asbestos is not –

(a) stored;

(b) received into or despatched from any place of work; or

(c) distributed within any place of work, except in a totally enclosed distribution system,

unless it is in a sealed receptacle or, where more appropriate, sealed wrapping, clearly marked in accordance with paragraphs (2) and (3) showing that it contains asbestos.

(2) Raw asbestos shall be labelled in accordance with the provisions of Schedule 2.

(3) Waste containing asbestos shall be labelled –

(a) where the Carriage of Dangerous Goods and Use of Transportable Pressure Equipment Regulations 2004[a] *apply, in accordance with those Regulations; and*

24 *(b) in any other case in accordance with the provisions of Schedule 2.*

(a) S.I. 2004/568, as amended by S.I. 2005/1732.

Guidance 24

356 Asbestos samples includes samples which are suspected of containing asbestos.

ACOP

Management of waste asbestos

357 Waste should be placed in suitable, labelled bags, wrapping or packaging as it is produced which, once sealed, ensures no asbestos fibres are released except that this shall not apply to intact asbestos cement sheeting in good condition and textured coatings firmly attached to a board. Where practicable, packaging should be sealed and the outside cleaned before removal from enclosures or the work area, and should be taken to a suitable and clearly identified secure storage area such as a lockable skip if they are not being disposed of at once.

358 Any waste where the escape of hazardous quantities of respirable asbestos fibres can occur during carriage should be placed in UN-approved packaging (available in up to 2 tonnes capacity) – this does not apply to asbestos cement or textured decorative coatings.

359 Bags, wrapping or packaging must be designed, constructed and maintained to ensure that no asbestos fibres can be released during normal handling. For most waste, double plastic sacks are suitable provided they will not split during normal use. It is important that the inner bag is not overfilled, especially when the debris is wet, and each bag should be capable of being securely tied or sealed. Air should be excluded from the bag as far as possible before sealing. Precautions will need to be taken as the exhaust air may be contaminated. Stronger packages are necessary if the waste contains sharp metal fragments or other materials liable to puncture plastic sacks.

24 **360 Wherever practicable, large pieces of rigid material, and in particular sheets of asbestos cement, must not be broken or cut for disposal in plastic sacks. They must be double-wrapped intact in suitable gauge polythene sheeting or other suitable material which must then be labelled. The wrapped and labelled material should then be placed in a sealed, labelled receptacle such as a lockable skip or freight container.**

Guidance 24

361 You may need a waste management licence from the relevant environment agency if you intend to sort ACMs from other debris or you want to re-use rubble contaminated with ACMs on the same site. *The licensed contractors' guide*[13] provides further advice on the disposal of waste.

ACOP

Transport of waste asbestos

362 Bags, wrapping or packaging containing asbestos waste should be appropriately labelled and transported to a licensed disposal site in an enclosed vehicle, skip or freight container. A receptacle should be used to transport the asbestos waste that ensures that the bags, wrapping and packaging cannot become damaged or otherwise opened in a way that may release asbestos material or asbestos fibres. The specific requirements of various Hazardous Waste Regulations in England and Wales and the Special Waste Regulations in Scotland should be adhered to, as appropriate.

Labelling of asbestos waste

363 Asbestos waste must be labelled:

24 **(a) in accordance with the Carriage of Dangerous Goods and Use of Transportable Pressure Equipment Regulations 2004[25] where those Regulations apply;**

ACOP
24

(b) where the Regulations in (a) do not apply, in accordance with Schedule 2 of the Asbestos Regulations.

Guidance
24

364 *The licensed contractors' guide*[13] contains more detailed advice on waste handling.

PART 3 PROHIBITIONS AND RELATED PROVISIONS

Regulation 25

Interpretation of prohibitions

Regulation

(1) In this Part –

"asbestos cement" means a material which is predominantly a mixture of cement and chrysotile and which when in a dry state absorbs less than 30% water by weight;

"asbestos spraying" means the application by spraying of any material containing asbestos to form a continuous surface coating;

"extraction of asbestos" means the extraction by mining or otherwise of asbestos as the primary product of such extraction, but shall not include extraction which produces asbestos as a by-product of the primary activity of extraction;

"supply" means supply by way of sale, lease, hire, hire-purchase, loan, gift or exchange for a consideration other than money, whether (in all cases) as principal or as agent for another; and

"use" in relation to asbestos or any product to which asbestos has intentionally been added means –

(a) putting asbestos or any product to which asbestos has intentionally been added to use for the first time; or

(b) putting asbestos or any product to which asbestos has intentionally been added which has been in use before to a new use.

(2) Any prohibition imposed on any person by this Part shall apply only to acts done in the course of a trade, business or other undertaking (whether for profit or not) carried on by him.

(3) Any prohibition imposed by this Part on the importation into the United Kingdom, or on the supply or use of asbestos shall not apply to the importation, supply or use of asbestos solely for the purposes of research, development or analysis.

(4) Where in this Part it is stated that asbestos has intentionally been added to a product or is intentionally added, it will be presumed where –

(a) asbestos is present in any product; and

(b) asbestos is not a naturally occurring impurity of that product, or of any component or constituent thereof,

25 *that the asbestos has intentionally been added or is intentionally added, as the case may be, subject to evidence to the contrary being adduced in any proceedings.*

Regulation 26

Prohibitions of exposure to asbestos

Regulation

(1) No person shall undertake asbestos spraying or working procedures that involve using low-density (less than 1g/cm^3) insulating or soundproofing materials which contain asbestos.

(2) Every employer shall ensure that no employees are exposed to asbestos during the extraction of asbestos.

(3) Every employer shall ensure that no employees are exposed to asbestos during the manufacture of asbestos products or of products containing intentionally added asbestos.

(4) In the case of chrysotile only, the prohibition in paragraph (3) is subject to the exception in paragraph 2 of Schedule 3.

26

Regulation 27

Prohibition of the importation of asbestos

Regulation

(1) Subject to paragraph (2), the importation into the United Kingdom of asbestos or of any product to which asbestos has intentionally been added is prohibited and any contravention of this paragraph shall be punishable under the Customs and Excise Management Act 1979[a] and not as a contravention of a health and safety regulation.

(2) In the case of chrysotile only, the prohibition in paragraph (1) is subject to the exceptions in paragraphs 1, 2 and 3 of Schedule 3.

27

(a) 1979 c.2.

Regulation 28

Prohibition of the supply of asbestos

Regulation

(1) Subject to paragraphs (2) and (3), no person shall supply, other than solely for the purpose of disposal, asbestos or any product to which asbestos has intentionally been added.

(2) In the case of chrysotile only, the prohibition in paragraph (1) shall not apply where the asbestos or the product was in use before 24th November 1999, except in the case of a product to which asbestos has intentionally been added of which the supply was prohibited by regulation 7 of the Asbestos (Prohibitions) Regulations 1992[a] as in force immediately before 24th November 1999.

(3) In the case of chrysotile only, the prohibition in paragraph (1) is subject to the exceptions in paragraphs 1 and 2 of Schedule 3.

28

(a) S.I. 1992/3067, as amended by S.I. 1999/2373.

Regulation 29

Prohibition of the use of asbestos

Regulation

(1) Subject to paragraphs (2) to (6), no person shall use, except in the course of any activity in connection with its disposal, asbestos or any product to which asbestos has intentionally been added.

(2) In the case of products containing crocidolite or asbestos grunerite (amosite), the prohibition in paragraph (1) shall not apply where the product was in use before 1st January 1986.

(3) In the case of products containing any other form of asbestos than crocidolite or asbestos grunerite (amosite), but excepting chrysotile, the prohibition in paragraph (1) shall not apply where the product was in use before 1st January 1993.

(4) In the case of chrysotile only, the prohibition in paragraph (1) shall not apply where the asbestos or product was in use before 24th November 1999, except in the case of a product containing chrysotile of which the supply was prohibited by regulation 7 of the Asbestos (Prohibitions) Regulations 1992 as in force immediately before 24th November 1999.

(5) Notwithstanding paragraph (4), no person shall use, except in the course of any activity in connection with its disposal –

(a) asbestos cement;

(b) any board, panel or tile, all or part of which has been painted with paint containing chrysotile; or

(c) any board, panel or tile, all or part of which has been covered in a textured finishing plaster used for decorative purposes and containing chrysotile,

unless it is installed in or forms part of any premises or plant and, before 24th November 1999, it was installed in or formed part of those same premises or plant.

29 *(6) In the case of chrysotile only, the prohibition in paragraph (1) is subject to the exceptions in paragraphs 1 and 2 of Schedule 3.*

Regulation 30

Labelling of products containing asbestos

Regulation

(1) Subject to paragraph (2), a person shall not supply under an exception in Schedule 3 or an exemption granted pursuant to regulation 32 or regulation 33 a product which contains asbestos unless that product is labelled in accordance with the provisions of Schedule 2.

(2) Where a component of a product contains asbestos, it shall be sufficient compliance with this regulation if that component is labelled in accordance with the provisions of Schedule 2 except that where the size of that component makes it impossible for a label to be fixed to it neither that component nor the product need be labelled.

30

Regulation 31

Additional provisions in the case of exceptions and exemptions

Regulation

(1) Where under an exception in Schedule 3 or an exemption granted pursuant to regulation 32 or regulation 33 asbestos is used in a work process or is produced by a work process, the employer shall ensure that the quantity of asbestos and materials containing asbestos at the premises where the work is carried out is reduced to as low a level as is reasonably practicable.

(2) Subject to paragraph (3), where under an exception in Schedule 3 or an exemption granted pursuant to regulation 32 or regulation 33 a manufacturing process which gives rise to asbestos dust is carried out in a building, the employer shall ensure that any part of the building in which the process is carried out is –

(a) so designed and constructed as to facilitate cleaning; and

(b) is equipped with an adequate and suitable vacuum cleaning system which shall, where reasonably practicable, be a fixed system.

(3) Paragraph 2(a) shall not apply to a building in which, prior to 1st March 1988, there was carried out a process to which either –

(a) as then in force, regulation 13 of the Asbestos Regulations 1969[a] *applied and the process was carried out in compliance with that regulation; or*

(b) that regulation did not apply.

31

(a) S.I. 1969/690–revoked by S.I. 1987/2115.

PART 4 MISCELLANEOUS

Regulation 32

Exemption certificates

Regulation

(1) Subject to paragraph (4), the Executive may, by a certificate in writing, exempt any person or class of persons or any product containing asbestos or class of such products from all or any of the requirements or prohibitions imposed by regulations 4, 8, 12, 13, 21, 22(5) to (7) and 27 and any such exemption may be granted subject to conditions and to a limit of time and may be varied or revoked by a further certificate in writing at any time.

(2) Subject to paragraph (4) and to the provisions of Council Directive 76/769/EEC on the marketing and use of certain dangerous substances and preparations,[a] *the Executive may, by a certificate in writing, exempt any person or class of persons or any product containing asbestos or class of such products from the prohibitions imposed by regulations 28(1) and 29(1) and any such exemption may be granted subject to conditions and to a limit of time and may be varied or revoked by a further certificate in writing at any time.*

(3) Subject to paragraph (4), the Executive may exempt emergency services from all or any of the requirements or prohibitions imposed by regulations 7 and 9 and any such exemption may be granted subject to conditions and to a limit of time and may be varied or revoked by a further certificate in writing at any time.

(4) The Executive shall not grant any exemption under paragraph (1), (2) or (3) unless having regard to the circumstances of the case and in particular to –

(a) the conditions, if any, which it proposes to attach to the exemption; and

(b) any other requirements imposed by or under any enactments which apply to the case,

it is satisfied that the health or safety of persons who are likely to be affected by the exemption will not be prejudiced in consequence of it.

(a) OJ No. L 262, 27.9.1976, p. 201, relevant amendments are Council Directives 83/478/EEC (OJ No L263, 24.9.83 p.33 and 85/467/EEC (OJ No L269, 11.10.85 p.56), and Commission Directives 91/659/EEC (OJ No L363, 31.12.91 p.36) and 1999/43/EC (OJ No L207, 6.8.99 p.18).

32

Regulation 33

Exemptions relating to the Ministry of Defence

Regulation

The Secretary of State for Defence may, in the interests of national security, exempt any person or class of persons from all or any of the prohibitions imposed by Part 3 of these Regulations by a certificate in writing, and any such exemption may be granted subject to conditions and to a limit of time and may be varied or revoked by a further certificate in writing at any time.

33

Regulation 34

Extension outside Great Britain

Regulation

These Regulations shall apply to any work outside Great Britain to which sections 1 to 59 and 80 to 82 of the Health and Safety at Work etc. Act 1974 apply by virtue of the Health and Safety at Work etc. Act 1974 (Application Outside Great Britain) Order 2001[a] as they apply to work in Great Britain.

34

(a) S.I. 2001/2127.

Regulation 35

Existing licences and exemption certificates

Regulation

(1) An existing licence issued by the Executive under regulation 4(1) of the Asbestos (Licensing) Regulations 1983[a] shall –

(a) continue to have effect as if it had been granted under regulation 8(2) of these Regulations;

(b) be of the duration and subject to the conditions specified in it as if that duration and those conditions had been specified under regulation 8(3); and

(c) be liable to variation and revocation under regulation 8(4) and (5),

and any requirement in such a licence concerning notification and any exception to such a requirement shall have effect as a requirement for notification under regulation 9 of and as an exception to such a requirement under regulation 3(2) of these Regulations.

(2) An existing exemption granted by the Executive under regulation 7(1) of the Asbestos (Licensing) Regulations 1983, regulation 8(1) of the Asbestos (Prohibitions) Regulations 1992,[b] or regulation 25(1) of the Control of Asbestos at Work Regulations 2002[c] shall continue to have effect and be subject to any limitation of time or any conditions specified in it and liable to revocation as if it had been granted under regulation 32(1), (2) or (3) of these Regulations.

(3) An existing exemption granted by the Secretary of State for Defence under regulation 8(3) of the Asbestos (Prohibitions) Regulations 1992 shall continue to have effect and be subject to any limitation of time or any conditions specified in it and liable to revocation as if it had been granted under regulation 33 of these Regulations.

35

(a) S.I. 1983/1649, as amended by S.I. 1986/392 and S.I. 1998/3233.
(b) S.I. 1992/3067, as amended by S.I. 1999/2373, S.I. 1999/2977 and S.I. 2003/1889.
(c) S.I. 2002/2675.

Regulation 36

Revocations, amendments and savings

Regulation

36

(1) The revocations listed in Schedule 4 shall have effect.

(2) The amendments listed in Schedule 5 shall have effect.

Regulation 36

(3) Any record or register required to be kept under any Regulations revoked either by paragraph (1) or by regulation 27(1) of the Control of Asbestos at Work Regulations 2002 shall, notwithstanding that revocation, be kept in the same manner and for the same period as specified in those Regulations as if these Regulations had not been made, except that the Executive may approve the keeping of records at a place or in a form other than at the place where, or in the form which, records were required to be kept under the Regulations so revoked.

Regulation 37 Defence

Regulation 37

Subject to regulation 21 of the Management of Health and Safety at Work Regulations 1999,[(a)] in any proceedings for an offence consisting of a contravention of Part 2 of these Regulations it shall be a defence for any person to prove that he took all reasonable precautions and exercised all due diligence to avoid the commission of that offence.

(a) S.I. 1999/3242, amended by S.I. 2003/2457 and S.I. 2006/438.

Schedule 1

Particulars to be included in a notification

Schedule 1

Regulation 9(1)

The following particulars are to be included in a notification made in accordance with regulation 9(1), namely –

(a) the name and address of the notifier and the address and telephone number of his usual place of business;

(b) a brief description of –

(i) the location of the work site,

(ii) the type(s) of asbestos to be used or handled (classified in accordance with regulation 2),

(iii) the maximum quantity of asbestos of each type to be held at any one time at the premises at which the work is to take place,

(iv) the activities and processes involved,

(v) the number of workers involved, and

(vi) the measures taken to limit the exposure of employees to asbestos, and

(c) the date of the commencement of the work and its expected duration.

Schedule 2

The labelling of raw asbestos, asbestos waste and products containing asbestos

Schedule

2

Regulations 14(4), 24(2) and (3) and 30(1) and (2)

1 *(1) Subject to sub-paragraphs (2) and (3) of this paragraph, the label to be used on –*

(a) raw asbestos (together with the labelling required under the Chemicals (Hazard Information and Packaging for Supply) Regulations 2002[(a)] and the Carriage of Dangerous Goods and Use of Transportable Pressure Equipment Regulations 2004);[(b)]

(b) asbestos waste, when required to be so labelled by regulation 24(3); and

(c) products containing asbestos, including used protective clothing to which regulation 14(2) applies,

shall be in the form and in the colours of the following diagram and shall comply with the specifications set out in paragraphs 2 and 3.

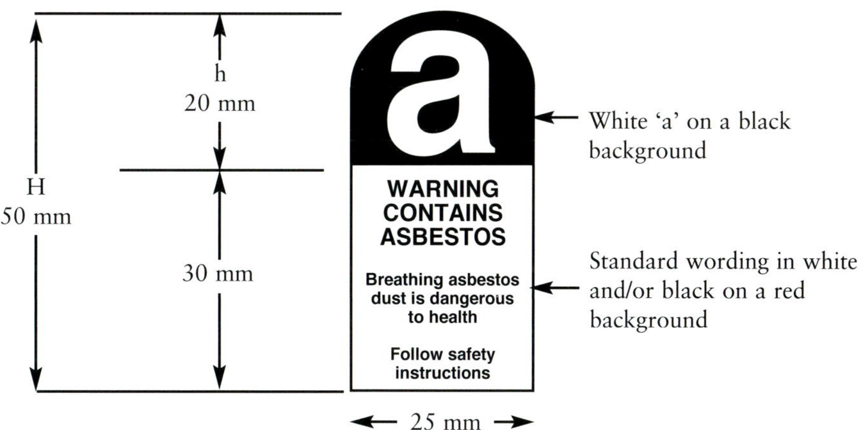

(2) In the case of a product containing crocidolite, the words "contains asbestos" shown in the diagram shall be replaced by the words "contains crocidolite/blue asbestos".

(3) Where the label is printed directly onto a product, a single colour contrasting with the background colour may be used.

2 *The dimensions in millimetres of the label referred to in paragraph 1(1) shall be those shown on the diagram in that paragraph, except that larger measurements may be used, but in that case the dimension indicated as h on the diagram shall be 40% of the dimension indicated as H.*

3 *The label shall be clearly and indelibly printed so that the words in the lower half of the label can be easily read, and those words shall be printed in black or white.*

(a) S.I. 2002/1689, as amended by S.I. 2004/568.
(b) S.I. 2004/568, as amended by S.I. 2005/1732.

Schedule 2

4 (1) Where a product containing asbestos may undergo processing or finishing it shall bear a label containing safety instructions appropriate to the particular product and in particular the following instructions –

"operate if possible out of doors in a well-ventilated place";

"preferably use hand tools or low speed tools equipped, if necessary, with an appropriate dust extraction facility. If high speed tools are used, they should always be so equipped";

"if possible, dampen before cutting or drilling"; and

"dampen dust, place it in a properly closed receptacle and dispose of it safely".

(2) Additional safety information given on a label shall not detract from or contradict the safety information given in accordance with sub-paragraph (1).

5 (1) Labelling of packaged and unpackaged products containing asbestos in accordance with the foregoing paragraphs shall be effected by means of –

(a) an adhesive label firmly affixed to the product or its packaging, as the case may be;

(b) a tie-on label firmly attached to the product or its packaging, as the case may be; or

(c) direct printing onto the product or its packaging, as the case may be.

(2) Where, in the case of an unpackaged product containing asbestos, it is not reasonably practicable to comply with the provisions of sub-paragraph (1) the label shall be printed on a suitable sheet accompanying the product.

(3) Labelling of raw asbestos and asbestos waste shall be effected in accordance with sub-paragraph (1)(a) or (c).

(4) For the purposes of this Schedule but subject to sub-paragraph (5), a product supplied in loose plastic or other similar wrapping (including plastic and paper bags) but no other packaging, shall be treated as being supplied in a package whether the product is placed in such wrapping at the time of its supply or was already so wrapped previously.

(5) No wrapping in which a product is placed at the time of its supply shall be regarded as packaging if any product contained in it is labelled in accordance with the requirements of this Schedule or any other packaging in which that product is contained is so labelled.

Schedule 3

Exceptions to the prohibitions on the importation, supply and use of chrysotile

Schedule

Regulations 26(4), 27(2), 28(3) and 29(6)

1 Regulations 27(1), 28(1) and 29(1) shall not apply to brake linings within the meaning of the Road Vehicles (Brake Linings Safety) Regulations 1999.[a]

2 Where it is not practicable for an employer to substitute for chrysotile a substance which, under the conditions of its use, does not create a risk to the health of his employees or creates a lesser risk than that created by chrysotile, regulations 26(3), 27(1), 28(1) and 29(1) shall not apply to –

(a) Diaphragms for use in electrolytic cells in existing electrolysis plants for chlor-alkali manufacture;

(b) chrysotile, or products to which chrysotile has intentionally been added, required solely for the manufacture of the products described in sub-paragraph (a).

3 Regulation 27(1) shall not apply to receptacles used for the storage of acetylene gas under pressure and in use before 24th November 1999.

3

(a) S.I. 1999/2978, as amended by S.I. 2003/3314.

Schedule 4

Revocations

Schedule

Regulation 36(1)

Instruments revoked	References	Extent of revocation
The Asbestos (Licensing) Regulations 1983	*S.I. 1983/1649*	*The whole Regulations*
The Personal Protective Equipment at Work Regulations 1992	*S.I. 1992/2966*	*Schedule 2 Part VII*
The Asbestos (Prohibitions) Regulations 1992	*S.I. 1992/3067*	*The whole Regulations*
The Asbestos (Licensing) (Amendment) Regulations 1998	*S.I. 1998/3233*	*The whole Regulations*
The Asbestos (Prohibitions) (Amendment) Regulations 1999	*S.I. 1999/2373*	*The whole Regulations*
The Asbestos (Prohibitions) (Amendment) (No. 2) Regulations 1999	*S.I. 1999/2977*	*The whole Regulations*
The Control of Asbestos at Work Regulations 2002	*S.I. 2002/2675*	*The whole Regulations*
The Asbestos (Prohibitions) (Amendment) Regulations 2003	*S.I. 2003/1889*	*The whole Regulations*
The Carriage of Dangerous Goods and Use of Transportable Pressure Equipment Regulations 2004	*S.I. 2004/568*	*Schedule 13 paragraph 12*
The Fire and Rescue Services Act 2004 (Consequential Amendments) (England) Order 2004	*S.I. 2004/3168*	*Article 63*
The Fire and Rescue Services Act 2004 (Consequential Amendments) (Wales) Order 2005	*S.I. 2005/2929*	*Article 73*
The Fire (Scotland) Act 2005 (Consequential Modifications and Amendments) (No 2) Order 2005	*S.S.I. 2005/344*	*Schedule 1 Part 1 paragraph 26*

4

Schedule 5

Amendments

Schedule

Regulation 36(2)

Instruments amended	References	Amendments to have effect
The Personal Protective Equipment at Work Regulations 1992	*S.I. 1992/2966*	*In regulation 3(3)(c) for the words "the Control of Asbestos at Work Regulations 1987" substitute "the Control of Asbestos Regulations 2006"*
The Health and Safety (Enforcing Authority) Regulations 1998	*S.I. 1998/494*	*In Schedule 2 paragraph 4A for the words "the Control of Asbestos at Work Regulations 2002" substitute "the Control of Asbestos Regulations 2006"*
The Provision and Use of Work Equipment Regulations 1998	*S.I. 1998/2306*	*In regulation 12(5)(b) for the reference to the Control of Asbestos at Work Regulations 1987 substitute a reference to these Regulations*
The Control of Substances Hazardous to Health Regulations 2002	*S.I. 2002/2677*	*In regulation 5(1)(a)(iii) for the words "the Control of Asbestos at Work Regulations 2002" substitute "the Control of Asbestos Regulations 2006"*
The Fur Farming (Compensation Scheme) (England) Order 2004	*S.I. 2004/1964*	*In Schedule 6 Part 6 paragraph 14(a)(i) and (ii) after the words "the Asbestos (Prohibitions) Regulations 1992" insert in each case the words "or, from 13th November 2006, by Part 3 of the Control of Asbestos Regulations 2006"*
The Health and Safety (Fees) Regulations 2006	*S.I. 2006/336*	*In the heading to regulation 5 for the words "the Asbestos (Licensing) Regulations 1983" substitute "regulation 8 of the Control of Asbestos Regulations 2006".* *In regulation 5(1) for the words "the Asbestos (Licensing) Regulations 1983*[a] *("the 1983 Regulations")" substitute "regulation 8 of the Control of Asbestos Regulations 2006 ("the 2006 Regulations")"* *In regulation 5(3) for the words "the 1983 Regulations" substitute "regulation 8 of the 2006 Regulations"* *In regulation 5(5) and (7) for the words "the 1983 Regulations" insert on each occasion the words "the Asbestos (Licensing) Regulations 1983 or regulation 8 of the 2006 Regulations"* *In regulation 6(3) for the words "the Control of Asbestos at Work Regulations 2002"*[b] *substitute "the Control of Asbestos Regulations 2006"*

5

(a) S.I. 1983/1649, as amended by S.I. 1986/392 and S.I. 1998/3233.
(b) S.I. 2002/2675.

Schedule

5

Instruments amended	References	Amendments to have effect
		In the heading to Schedule 4 for the words "the Asbestos (Licensing) Regulations 1983" substitute "regulation 8 of the Control of Asbestos Regulations 2006".
		In Schedule 4 Table 1 column 1 after the words "work with asbestos" delete the words "insulation or asbestos coating or asbestos insulating board"
		In Schedule 5 row (b) columns 1 and 2 for the reference to the Control of Asbestos at Work Regulations 2002 substitute a reference to these Regulations
The Health and Safety (Enforcing Authority for Railways and Other Guided Transport Systems) Regulations 2006	*S.I. 2006/557*	*In regulation 4(5) for the words "the Asbestos (Licensing) Regulations 1983" substitute "the Control of Asbestos Regulations 2006" and for the words "regulation 4" substitute "regulation 8"*

References

1 *Control of Asbestos Regulations 2006* SI 2006/2739 The Stationery Office 2006 ISBN 0 11 075191 4

2 *The management of asbestos in non-domestic premises. Regulation 4 of the Control of Asbestos at Work Regulations 2006. Approved Code of Practice and guidance* L127 (Second edition) HSE Books 2006 ISBN 0 7176 6209 8

3 *Health and Safety at Work etc Act 1974 (c.37)* The Stationery Office 1974 ISBN 0 10 543774 3

4 *Safety representatives and safety committees* L87 (Third edition) HSE Books 1996 ISBN 0 7176 1220 1

5 *A guide to the Health and Safety (Consultation with Employees) Regulations 1996. Guidance on Regulations* L95 HSE Books 1996 ISBN 0 7176 1234 1

6 *Consulting employees on health and safety: A guide to the law* Leaflet INDG232 HSE Books 1996 (single copy free or priced packs of 15 ISBN 0 7176 1615 0) Web version: hse.gov.uk/pubns/indg232.pdf

7 *Managing health and safety in construction: Construction (Design and Management) Regulations 1994. Approved Code of Practice and guidance* HSG224 HSE Books 2001 ISBN 0 7176 2139 1

8 *Asbestos: The analysts' guide for sampling, analysis and clearance procedures* HSG248 HSE Books 2005 ISBN 0 7176 2875 2

9 BS EN 60335 *Specification for safety of household and similar electrical appliances*

10 *Asbestos essentials task manual: Task guidance sheets for the building maintenance and allied trades* HSG210 HSE Books 2001 ISBN 0 7176 1887 0

11 *Surveying, sampling and assessment of asbestos-containing materials* MDHS100 HSE Books 2001 ISBN 0 7176 2076 X Web version: www.hse.gov.uk/pubns/mdhs/index.htm

12 *Management of health and safety at work. Management of Health and Safety at Work Regulations 1999. Approved Code of Practice and guidance* L21 (Second edition) HSE Books 2000 ISBN 0 7176 2488 9

13 *Asbestos: The licensed contractors' guide* HSG247 HSE Books 2006 ISBN 0 7176 2874 4

14 *Working with asbestos cement* HSG189/2 HSE Books 1999 ISBN 0 7176 1667 3

15 *Health and safety in roof work* HSG33 (Second edition) HSE Books 1998 ISBN 0 7176 1425 5

16 *Control of substances hazardous to health (Fifth edition). The Control of Substances Hazardous to Health Regulations 2002 (as amended). Approved Code of Practice and guidance* L5 (Fifth edition) HSE Books 2005 ISBN 0 7176 2981 3

17 PAS 60-1:2004 *Equipment used in the controlled removal of asbestos-containing materials. Controlled wetting of asbestos-containing materials. Specification* British Standards Institution

18 *Fit testing of respiratory protective equipment facepieces* OC 282/28 available to view online at: www.hse.gov.uk/pubns/fittesting.pdf

19 PAS 60-3:2004 *Equipment used in the controlled removal of asbestos-containing materials. Operation, cleaning and maintenance of class H vacuum cleaners. Code of practice* British Standards Institution

20 PAS 60-2:2004 *Equipment used in the controlled removal of asbestos-containing materials. Negative Pressure Units. Specification* British Standards Institution

21 BS EN ISO 13982-1:2004 *Protective clothing for use against solid particulates. Performance requirements for chemical protective clothing providing protection to the full body against airborne solid particulates (type 5 clothing)* British Standards Institution

22 BS EN ISO/IEC 17020:2004 *General criteria for the operation of various types of bodies performing inspection* British Standards Institution

23 BS EN ISO/IEC 17025:2005 *General requirements for the competence of testing and calibration laboratories* British Standards Institution

24 *Safe work in confined spaces. Confined Spaces Regulations 1997. Approved Code of Practice, Regulations and guidance* L101 HSE Books 1997 ISBN 0 7176 1405 0

25 *Carriage of Dangerous Goods and Use of Transportable Pressure Equipment Regulations 2004* SI 2004 No 568 The Stationery Office ISBN 0 11 049063 0